The Vault of Walt: Volume 4

Books by Jim Korkis

The Vault of Walt: Volume 4 (Theme Park Press, 2015)

Everything I Know I Learned from
Disney Animated Features (Theme Park Press, 2015)

The Vault of Walt: Volume 3 (Theme Park Press, 2014)

Animation Anecdotes (Theme Park Press, 2014)

Who's the Leader of the Club?
Walt Disney's Leadership Lessons (Theme Park Press, 2014)

The Book of Mouse (Theme Park Press, 2013)

The Vault of Walt: Volume 2 (Theme Park Press, 2013)

Who's Afraid of the Song of the South? (Theme Park Press, 2012)

The Revised Vault of Walt (Theme Park Press, 2012)

The Vault of Walt: Volume 4

Still More Unofficial Disney Stories Never Told

Jim Korkis

Foreword by Jeff Kurtti

Theme Park Press
www.ThemeParkPress.com

Theme Park Press publishes its books in a variety of print and electronic formats. Some content that appears in one format may not appear in another.

Editor: Bob McLain
Layout: Artisanal Text

ISBN 978-1-941500-62-0
Printed in the United States of America

Theme Park Press | www.ThemeParkPress.com
Address queries to bob@themeparkpress.com

This book is dedicated to Didier Ghez, an outstanding friend and fellow Disney historian, whose persistence, patience, generosity, and enthusiasm enriches Disney scholarship and inspires so many other people.

Didier is responsible for many great books sharing Disney history, including the critically praised *Walt's People* series available from Theme Park Press. Without him, many Disney stories, including some in this book, would have been lost forever.

Contents

PART FOUR: Other Disney Stories

Foreword

In the age of the internet, "Disney historians" are a dime a dozen. There appears to be a conventional wisdom that if you like something, and can cobble some cohesive sentences together, and post them for people to read, some degree of expertise is inherent, or at least implied.

Nothing could be further from the truth. I recently spent far too many hours on a web site trying to pull the plug on yet another Disney myth being put forward as "truth" by an overly enthusiastic—and quite ill-informed—fan. (Seriously, his resource was that "a bartender at [the location being discussed] told me." So much for scholarship.)

In their eagerness to be a part of the community, too many people don't start from the beginning. It's important to learn from those who have gone before. Bob Thomas. Pete Martin. Christopher Finch. Frank Thomas and Ollie Johnston. John Canemaker. Brian Sibley. J.B. Kaufman. The collected works of Didier Ghez and his colleagues. And as many eyewitnesses as might be left.

Information needs to be weighed in an overall balance, contextualized, and perpetually cross-referenced, checked, and confirmed (or denied). Too much scholarship is left by the wayside as people rush to publish without seeking proper sources. Even eyewitness accounts are often questionable, altered by time, "enhanced" for retelling, and requiring context and corroboration. "That's one of the reasons historians have to be careful," Jim Korkis says, "even if they are getting the information from a first-hand source."

As a writer, documentarian, curator, lecturer, consultant, Disney historian, and overall fan, I'm delighted to tell you that what you hold is "the genuine article". I have known Jim Korkis for years (decades?) since our first collaborations during the glory days of The Disney Institute at Walt Disney World.

Over the years I have relied on Jim as a resource, researcher, sounding board, and contributor to many projects, including the member newsletter and blog for The Walt Disney Family Museum in San Francisco. Jim balances deep and informed research and investigation with a proper amount of skepticism and questioning. Jim has certitude where appropriate, and doubt where necessary. He is a confident historian, and a fearless yet humble enquirer.

It's a boon to any Disney fan of any level of interest to have another collection of Jim's excellent essays to call on. The stories here are varied and fascinating, and shed light on frequently unknown or unconnected bits of Disney history, or illuminate the character and personality of Walt Disney himself. There is a wonderful alchemy that results from Jim's combination of scrupulous research, informed passion, and the felicity of well-turned storytelling.

As the years move by, so many of these stories, and the thoughts and memories of the people who tell them, would have been lost to time. In addition to the value of preserving this information, Jim has done so in a way that educates and entertains his readers, and I believe not only inspires their own scholarship, but also informs how they seek their information—and how much they trust their sources—and challenges those who wish to follow in his footsteps as historians.

So it's time again to open up *The Vault of Walt* and enjoy the work of Walt Disney, his colleagues, and creative heirs—presented by someone who carries on a tradition of quality and value commensurate with his subject matter.

Enjoy!

Jeff Kurtti

Jeff Kurtti is one of the leading authorities on The Walt Disney Company and its history. The author of more than 25 books, Kurtti worked for Walt Disney Imagineering, the theme park design division of the company, and then for Disney's Corporate Special Projects Department. He was creative director, content consultant, and media producer for The Walt Disney Family Museum.

Introduction

Is there anything new still left to be said about anything in Disney history?

There are a plethora of books, websites, blogs, and podcasts devoted to sharing stories of Disney history.

In fact, most of them simply repeat the same stories and "secrets" over and over, indicating that all there is to know about Disney is already known, or will never be known since the people involved have passed away and documentation has disappeared.

Ironically, there is more to be discovered and shared about Disney today than at any other time in recorded history.

Several unpublished memoirs and scrapbooks have been recently uncovered and put into print. Thanks to the internet, people now have access to international sources as well as newspaper archives, previously out-of-print material, and even direct contact with people who worked at Disney.

Countless books, especially from Theme Park Press, have unlocked previously unknown stories and insights, and many more are slated to be published in the near future.

When it came time to put together this latest volume in the Vault of Walt series, I worried that I might not have enough good and rarely shared stories, especially compared to the previous editions. That was not the problem.

The problem was that I had too many good stories and I had to decide which ones to leave on the cutting room floor, perhaps for a future edition. The original table of contents changed several times from first submission to the publisher to final proof copy.

Like the previous editions in this series, I think I can guarantee that these are Disney stories that you have not heard before, or at least not with this amount of detail and documentation.

Of course, new information is always appearing. Everyone "knew" that Walt Disney must have climbed into the cabin of a train on the Santa Fe & Disneyland Railroad to take over as engineer every now and then, but there was no proof.

Yet, just last year, an old Retlaw Enterprises (Walt's private company that owned the trains and the use of his name) itinerary kept by a former employee who was getting rid of it clearly showed that every Sunday in 1957 at 2:00 pm, Walt would tap out the engineer and, dressed in his overalls and cap, take the train around the park for a few hours without the guests ever knowing.

Everyone "knew" for decades that Walt Disney would have never flashed a "Vote for Goldwater" button to Goldwater's opponent in the upcoming election, President Lyndon Johnson, on September 14, 1963, when Johnson awarded Walt the nation's highest civilian honor, the prestigious Medal of Freedom. And, yet, that is pretty much what Walt did do and the full facts behind it were unearthed just a few years ago by Disney historian and author Michael Barrier.

Disney historian Didier Ghez's recent book, *Disney's Grand Tour: Walt and Roy's European Vacation, Summer 1935,* revealed new information and a new perspective on that historic trip as well as debunking Walt's meeting with Italian dictator Mussolini that everyone "knew" must have taken place. It didn't.

Those are just some of the reasons that it is exciting to be researching and writing Disney history today. New information and new perspectives are constantly being shared. For those people who are willing to put in the long hours and many frustrations, precious gems can still be found.

A handful of those previously unshared gems are inside these pages, and I sincerely hope that they will inspire others to go out and find even more information.

Disney history is such a rich and diverse tapestry that some of the little details are too often overlooked and are in danger of being forgotten. The Vault of Walt series is my attempt to showcase some of them for your enjoyment and to help future researchers.

Thanks for buying this book.

Jim Korkis
July 2015

PART ONE
Walt Disney Stories

George Washington spent many nights in inns and private homes as a surveyor, then a colonial officer and finally as the first president of the United States.

It was once a cliché for realtors attempting to sell an antiquated dwelling on the eastern seaboard of the United States to proudly proclaim, "George Washington slept here." That phrase was so well worn that it also became the title of a 1940 stage (and later movie) comedy written by Kaufman and Hart that kidded the concept.

Many of those claims were verified by public records or written family documents like personal correspondence from the time, but just as many were tall tales adding a little more prestige to the property.

People knew that even today others were greatly intrigued to be somewhere that had been intimately touched by a person of greatness.

Walt Disney World has four theme parks, multiple themed resorts, two water parks, and a downtown shopping area, so there is greater value spending time there. Yet, many people always choose to go to the smaller Disneyland with its fewer offerings instead.

The explanation, according to the many surveys over the years taken by the Disney Company, is that "this is where Walt Disney actually walked". Walt had a private apartment above the firehouse on Main Street and was often known to wander the park late at night or early in the morning before paying guests swarmed in.

Even with all the extreme changes that have happened to Disneyland after Walt passed away nearly half a century ago, the idea that this is where Walt "lived" holds an emotional grip on not only guests, but the cast members who work there.

That idea is reinforced by the common perception that spirits often remain in a physical location that they loved dearly and about which they hold many fond memories.

Even today, Disneyland keeps the lamp in the window of Walt's apartment constantly lit. When Walt was staying at the park, if the lamp was on, it let employees know that Walt was physically there. So Disneyland Guest Relations will tell guests that the lamp is always lit to remind people that Walt is still there, in spirit.

If Walt's spirit still lingers at the Happiest Place on Earth, he may actually be a hitchhiking ghost following some of the departing guests to other locations in the Los Angeles area that also hold many fond memories for him.

Walt was not a hermit. He could be found at the Hollywood Park race track betting on the horses. He could be found at a nearby sports stadium watching a baseball game. He could even be found occasionally on the grounds of Forest Lawn Memorial Park walking with his mom and dad who enjoyed the beauty of the place.

In the following chapters, I have concentrated on the most significant locations where Walt lived, worked. and visited.

Many years ago, an avid Disney fan and artist who had worked for the Disney Company for many years went to the birthplace of Walt Disney at 2156 North Tripp Avenue in Chicago, Illinois.

At the time, the house was under private ownership. With a huge smile, he eagerly bounded up the path and knocked on the door. An older woman answered.

"Do you know whose house this is?" he gleefully asked, hoping to surprise her with the news that she was inhabiting a landmark location.

"Yes," replied the woman tersely. "Mine!"

She then proceeded to turn on the sprinklers to drive him away and slammed the door.

Apparently, she had had just one too many Disney fans who wanted to genuflect at the site.

Keeping that story in mind, it is important to remember that many of the following locations are private property and not tourist attractions. If you visit, please be courteous and follow all restrictions.

All of the places have changed significantly since Walt was there, but a couple still have some of the "feel" of what they were like when he was alive and living in the Los Angeles area.

I have even included several places that were important to Walt, but no longer exist. However, there are a few where the owners can legitimately claim, "Walt Disney slept here."

Walt's Hollywood Homes

Walt Disney arrived in the Los Angeles area in August 1923. For over forty years, he continued to reside in the same general area until his death in December 1966. His homes were visibly less ostentatious than other movie studio heads and often included unique touches designed by Walt himself.

Uncle Robert Disney's Home
4406 Kingswell Avenue, Los Angeles CA 90027

This is the Los Feliz location where Walt's uncle, Robert Samuel Disney (the younger brother of Walt's father, Elias) lived. Walt stayed here in August and September of 1923 as he searched for work in Hollywood. His uncle charged Walt $5 a week rent to stay there, and it was often paid by Walt's older brother Roy as a "charity loan".

On the left hand side of the house was a small wooden garage (since relocated to Garden Grove) that Walt used as his first studio.

Walt and Roy lived briefly in a one-room apartment at Olive Hill in October and November.

In December 1923, Walt and Roy shared a room together at 4409 Kingswell in a house owned by Charles Schneider. It was directly across the street from Uncle Robert, and the room was $15 a month. After a year, the brothers had gotten on each other's nerves. Roy decided the best way to get a place of his own was to ask his fiancé to come to Los Angeles.

Roy and Edna Disney were married at Uncle Robert's house on April 11, 1925. Walt was the best man and Lillian Bounds was the maid of honor.

Walt did not like living alone and he got married to Lillian on July 13, 1925, in Lewiston, Idaho. The newlyweds rented a one-room apartment at 4637 North Melbourne Avenue (one street up from Kingswell), but it was much too small and later the couple moved to 1307 North Commonwealth Avenue. The entire 1300 and 1400 blocks of North Commonwealth Avenue no longer exist.

Walt Disney's First House
2495 Lyric Avenue, Los Angeles, CA 90027

In June 1926, with the money coming in from their new cartoon studio, Walt and Roy put a $200 deposit down on adjacent lots (Roy and Edna lived at 2491 Lyric Avenue that was next door to Walt and Lillian) at the bottom of the Los Feliz hills near their studio.

By August, they had purchased those lots for $1,000 dollars each. Walt's lot was a corner lot and was roughly 2,875 square feet. "We built two houses. They were the ready-cut [prefabricated] type of houses," Roy later told an interviewer. "The lot and the houses cost us $16,000." (That price was $8,000 for each house, including the lot. The construction was completed in December 1926.)

The Disney brothers had both purchased Pacific Ready-Cut homes for those lots. These were ready-to-assemble and shipped to the site, complete with knotless Douglas Fir framing, cabinets, nails, doors, windows, screens, hardware, paint, sinks, and an instruction manual. Altogether, it totaled approximately 12,000 pieces. Eventually, Pacific Ready-Cut sold more than 37,000 houses in southern California.

The interior of the prefabricated homes was small, less than 1100 square feet, with only two bedrooms, a living room, bath, dining room, and kitchen. The two houses were mirror images of each other.

Shortly after they moved in, Walt had Lillian's mother also move in to keep Lillian company while he worked long hours at the studio.

It was at this home, in the garage, that work was done (like painting of cels) on the first Mickey Mouse cartoon (*Plane Crazy* 1928) to avoid prying eyes at the Disney Studio. The house was denied historic status on July 21, 2000.

Walt and Lillian lived in the home from 1927 to just before the birth of their first child in 1933 when they moved to Woking Way.

In 1934, Roy and Edna moved out of their Lyric Avenue home and relocated to 4365 Forman Avenue in the North Hollywood/Burbank area.

In 1997, the interior of Walt's corner house was in such disrepair that the owner had it gutted to bare studs, with new electrical and plumbing installed throughout the house, although the exterior remained similar to the original.

Walt Disney's Second House
4053 Woking Way, Los Angeles, CA 90027

This $50,000 twelve-room French-Norman-style house was built in the summer of 1932 in a mere two and a half months to be ready in time for the arrival of Walt and Lillian's first child. Walt designed the home with architect Frank Crowhurst, who worked on a tower addition to Disney's Hyperion studio.

The primary work force was out-of-work day labor construction workers who were happy to find any type of temporary job during the Depression. The situation was comparable to people today picking up freelance handymen outside of a Home Depot or a Lowe's to help on a home project, except that according to Walt, the workmen showed up at the actual site each morning hoping to be selected for a few hours of paid work.

"We had been living in a little place where I couldn't turn around," Walt told an artist at the studio in 1944. "So I made the architect add three or four yards to every room in the house."

Unfortunately, Lillian miscarried. However, the Disneys were blessed with a daughter, Diane, slightly more a year later, in December 1933.

The house was about 6,300 square feet and was originally on almost 1.5 acres. It had a broad lawn that went down a hilly incline to the street. The location had winding, narrow, and sometimes steep streets.

Walt told newspaper columnist Hedda Hopper in 1964: "I found a graduate of the Vienna Academy of Fine Arts and had him paint my whole ceiling!"

Walt had a pool installed where he personally taught his daughters how to swim and often invited animators from his studio to come over and use it. "He hung this swimming pool up on the corner of this darn thing," recalled Roy in 1968. "It's a granite hill and we were taking bets to see if it would stand. It's 35 years and it's still there."

Walt commuted to the new Disney Studio in Burbank when it opened several years later by taking Riverside Drive around Griffith Park.

He had one of the bedrooms converted into a screening room primarily to view dailies (film footage shot on a particular day for later review) from his first live-action film, *Song of the South* (1946).

Walt's older daughter, Diane Disney Miller, told me:

> The making of this picture was the reason for the conversion of the downstairs guest room and bath-library wing to a projection room and small wet bar. Dad wanted to be able to watch the dailies at home.

However, the room was also used, like many celebrity home screening rooms, to watch popular movies of the day borrowed from other studios without having to go out to a regular movie theater.

Diane said that the interior Juliet balcony was named by her and her sister as "Christmas Tree Point" because on Christmas morning they opened the doors of their bedroom and stood there looking into the two-story living room with vaulted beamed ceiling and saw the huge decorated Christmas tree and all the presents beneath it.

She also remembered a special memorable gift from Santa when she was in elementary school:

> One Christmas, Santa Claus brought us [Diane and her younger sister, Sharon] a playhouse and I just knew that Santa Claus did because it just appeared Christmas morning out in our backyard. It was this darling little playhouse.
>
> It was designed at the [Disney] Studio and the studio carpenters put it up. It was a little one room, about the size of a good-sized closet. It had little leaded glass windows and one of those little mushroom chimneys on it, though there was no fireplace, and a sink with running water.
>
> It had a little tank inside the cooler that you filled then you could turn on the faucet and the water would come out. It had a little cooler all stocked with little tiny canned goods. You know, the small cans that you can buy. It had a telephone in it that would connect with our phone in the kitchen.

Lillian Disney, in the *McCall's* magazine article "I Live With a Genius" (February 1953), also recalled that little dwarf cottage:

> After *Snow White* came out [in 1937], it was so successful we felt flush about buying presents for the kids. The studio carpenters had spent days building a replica of the dwarfs' house for them. Then they [Diane and Sharon] started playing train with the boxes the things had come in.

The exterior of that cottage still exists today, although the original interior was gutted long ago.

Walt and Lillian socialized with actor Spencer Tracy and his wife at the Woking Way house for afternoons of swimming and badminton. Interestingly, these invitations were sent by letter or telegram, and never by phone. The house was featured in the January 1940 issue of *Better Homes and Gardens* magazine.

During this time Walt's parents, Elias and Flora, moved from Oregon to Los Angeles, and briefly lived in a rented apartment on Commonwealth Street until Walt and Roy moved them into a new home at 4605 Placedia in North Hollywood near where Roy lived. This is the house where Flora died of carbon monoxide poisoning from a faulty heater in November 1938.

(In an interesting bit of historical trivia, Leno and Rosemary LaBianca, both victims of the Charles Manson murders in 1969, had lived at the

Woking Way house shortly before the killings, and Rosemary's driver's license still listed that address.)

The Disney family moved out in 1950 and into a new home in Holmby Hills.

Over the years part of the Woking Way property was sold, including the section with the original pool that is now part of another property. The pool that exists today near the house was built in 1963. The interior of the house itself has also undergone renovations, such as changing Walt's workout room into a nursery and eventually a billiards room.

Russian filmmaker Timur Bekmambetov (*Abraham Lincoln: Vampire Hunter*; *Wanted*) is the current owner of the house.

"This is an iconic house. It should be treated like a museum," Bekmambetov told the media.

Bekmambetov, a Disney fan, is committed to preserving the memory of Walt. Disney artwork, historic photos and other artifacts decorate the rooms. He has opened up the home to private events, some from the American Film Institute and the Disney Studio (including a press day connected to the BluRay releases of *Maleficent* and *Sleeping Beauty*).

Walt Disney's Last House
355 Carolwood Drive, Los Angeles, CA 90077

With the Disney daughters becoming teenagers, Walt and Lillian decided they needed more room and joked that this new home was their 25th anniversary present to themselves. They officially moved in February 1950.

Lillian Disney telephoned Harold Janss about purchasing a parcel of property in the new subdivision that he was developing called Holmby Hills. After viewing the location, the Disneys, on June 1, 1948, acquired a parcel of land on which to build their dream home, and it took well over a year to complete.

Many celebrities lived in the Holmby Hills area over the years, including Frank Sinatra, Marilyn Monroe, Elvis Presley, Michael Jackson, and many others. The Disney parcel was on a knoll between Beverly Hills and Bel Air, with a view of UCLA.

Architect John Dolena designed the two-story, split-level main house of 5,669 square feet. The home flared out into two wings on either side, sometimes described as a "horseshoe" configuration.

It had seventeen rooms. However, it was not really palatial, especially when compared to the homes of other studio moguls and even other residents of the subdivision. In fact, Walt had assisted with the design so that the rooms would be easier to clean and would be practical for his growing family. In addition, Walt built a 22' by 45' swimming pool.

Beyond the pool was a two-story building, featuring a 1,566 square foot recreation room with motion picture projection equipment (so it served as a screening room), as well as a fully equipped ice cream soda fountain and liquor bar. Below the recreation room was a four-car garage and service area.

One of the reasons for Walt selecting this particular lot was that he determined there was enough land for his soon-to-be-famous miniature railroad, the Carolwood Pacific, named after the nearby street. The measurements for the right of way and trackage were done by architect John Cowles, Jr., who was also responsible for the design of the red barn on the property that Walt used as his workshop and control room for the railroad.

The railroad ran from December 1950 to spring 1953, when an accident involving a young girl being burned by steam from the engine resulted in Walt removing it from operation. "I got the power company and paid them a good price to remove or build a new power line behind me," Walt said. He wanted to make sure that no electrical lines could be seen from the railroad.

Landscaping for the estate (and the railroad) was done by Jack Evans and his younger brother Bill, whose work so pleased Walt that he hired them to landscape Disneyland.

Both Walt and Lillian continued to live in the house until their respective deaths in 1966 and 1997.

Gabriel Brener, chief executive of private investment firm Brener International Group and co-owner of the Houston Dynamo soccer team, purchased the property from the Disney estate for $8.45 million in 1998, a year after Lillian Disney's death.

Brener razed the original house, telling the media that there was structural damage and a plethora of asbestos so the house could not be saved, and erected a brand new 35,000 square-foot mansion in 2001. He also acquired the lot next door, adding more acreage.

He did keep the original gate, some of Lillian's rose garden and the tunnel for Walt's miniature Carolwood Pacific railroad which had been buried and decorated with landscaping. The entrance was marked by an ivy-covered miniature stone archway with the date "1950" (the year the railroad officially began operating) on it.

Walt's barn and the track were relocated to Griffith Park. Recognizing the historical and emotional importance of the barn, Diane Disney Miller began the process of rescuing it before escrow closed.

With the help of Michael and Sharon Broggie, founders of the Carolwood Pacific Historical Society, it officially opened July 19,1999, as part of the Los Angeles Live Steamers Museum in Griffith Park. It is open to the public the third Sunday of every month, or by special arrangement.

The Carolwood Estate came to market quietly in October of 2012 with an asking price of $90 million and eventually sold for $74 million in June 2014.

Walt's Hollywood Studios

Unable to find any type of work despite all his efforts at the local movie studios, Walt turned once again to doing animation as he had done in Kansas City, Missouri. He started small and with each new success expanded his operation until finally building the Disney studio that is used as the company's corporate offices today.

Uncle Robert Disney's Garage
4406 Kingswell Avenue, Los Angeles, CA 90027

In September 1923, Walt was living with his Uncle Robert (the younger brother of Walt's father, Elias) and was unable to find any type of work at the Hollywood live action motion picture studios. Frustrated, Walt decided to try to get back into doing animation. He bought an old used camera that was not in the best of shape for $200 from a local Los Angeles camera shop.

His older brother Roy had given Walt $10 to make up some business cards and letterhead paper proclaiming "Walt Disney, Cartoonist" and using Uncle Robert's address as the location for his art studio.

Walt asked Uncle Robert if he could set up his studio in the garage adjoining the house. Robert charged Walt an additional one dollar a week to use the garage, in addition to the five dollar a week rent Walt was already paying.

Walt had to tear up dry-goods boxes and find spare lumber to build a crude camera stand and animation set-up. The equipment would not accommodate anything more complicated than the simplest of animation.

Walt went to see Alexander Pantages, who owned a theater named after him, the Pantages, in downtown Los Angeles, as well as several other theaters. He pitched the idea of doing a series of short 30-second joke reels, like the Laugh-O-grams he had done for the Newman theaters in Kansas City. These reels would be weekly "exclusives" to run during the newsreel that would help publicize Pantages' theaters.

Pantages showed interest and Walt proceeded to film a sample reel using stick figures. Just before he could show the reel to Pantages, Walt received word that New York film distributor Margaret Winkler was ready to make a deal for Walt to produce a series of Alice Comedies combining a live-action little girl with animation.

Since the garage was never identified as the Walt Disney cartoon studio, nor had any anything produced there been sold, the Disney Company does not consider it the first Disney studio. The company sometimes officially identifies it as Walt's first Hollywood "workshop".

However, some early official Disney publications do identify the garage as Walt's first studio because it makes for a great "rags-to-riches" story.

The garage remained untouched for decades and was saved from destruction by a group called the Friends of Walt Disney in 1982, who donated it to the Stanley Ranch Museum (Heritage Park) located at 12174 Euclid Street, Garden Grove, California. (For the full story, pick up *Walt Disney's Garage of Dreams*, from Theme Park Press.)

In 2014, there was a Cadillac CTS Sedan television commercial titled "Garages". Narrated by Neal McDonough, it made the point that many business empires, including Hewlett-Packard, Amazon, Mattel, the Wright Brothers...and even Disney...all started in a garage.

The house used in the ad is located about a mile north of Robert Disney's actual home, but it is also a Los Feliz dwelling at 2223 Nella Vista Avenue. The plank wood doors of the real garage structure were positioned on the left-hand side, while the doors on the garage in the commercial are on the right. They did use the right street address for the house, 4406.

Walt Disney's First Studio
4651 Kingswell Avenue, Los Angeles, CA 90027

Since he needed a larger studio space to fulfill the work needed for the Alice Comedies, Walt walked down the street two blocks west from Uncle Robert's house to the Holly-Vermont Realty office.

As former head Disney Archivist Dave Smith wrote in March 1982:

> [Walt] did some preliminary work on a reel, probably in the garage, but word came on the purchase of a series of Alice Comedies, so the Pantages reel was never finished.

> Walt moved down the street on October 8, 1923, to 4651 Kingswell, and there in the back of a real estate office set up the first Disney Studio. A contract was signed for the Alice Comedies on October 16, 1923, the official date of the beginning of the Disney Studio.

Walt told the owners of the office that he only needed enough room "to swing a cat in" and could only afford a maximum of ten dollars a month. He was given a room at the back of the real estate office. Walt and Roy had to put up a curtain to block the room from the activity in the rest of the office.

In 1924, for the early Alice Comedies, Walt and Roy also rented a vacant lot (three blocks away) at the corner of Rodney Drive and Hollywood

Boulevard for ten dollars a month for shooting outdoor live-action segments for the series.

It was at this studio that Walt made the first Alice Comedy in Hollywood, *Alice's Day at Sea*, that distributor Margaret Winkler received December 26, 1923, the day after Christmas. Walt was the only animator until Rollin "Ham" Hamilton was hired to assist in February 1924. Lillian Bounds had been hired on January 19, 1924, to ink and paint.

Also in February 1924 for forty-two dollars a month, the Disneys expanded their studio into the empty storefront next to the real estate office. The address was 4649 Kingswell Avenue, and on the plate glass window was emblazoned in gold leaf the name "Disney Bros. Studio".

Roy's desk was at the front and could be seen through the window. To the left of the entrance were the ink and paint girls. At the rear was the camera room next to the room with the animation desks where Walt worked until they could hire an animation staff.

Today, the building is the home to Extra Copy, a photocopy business that has been there for many years. They have a homemade mural on the wall, as well as a bulletin board filled with clippings and photos, including a copy of Walt's death certificate. Enthusiastic employees (including the owner) love talking to tourists about the fact that it was Walt's first studio.

Walt's Hyperion Studio
2719 Hyperion Avenue, Los Angeles, CA 90029

As the success of the Disney Bros. Studio grew, Walt and Roy realized they needed their own studio not only to handle the current work, but to accommodate the expansions they wanted.

In 1925, a week before Walt's marriage to Lillian in July, Walt and Roy placed a $400 down payment on a plot of land between Griffith Park Boulevard and Monon Street on Hyperion Avenue.

Walt had let Roy make the choice of the Hyperion location over another plot in Westwood. There was an existing stucco building about 1,600 square feet, and the Disney brothers paid roughly $3,000 dollars for renovations.

Roy supposedly told Disney Legend Jack Cutting that Walt insisted the new studio be called Walt Disney Studios, although there are other versions of this story, including one Dave Smith heard from Roy that it was Roy's idea to change the name from the Disney Bros. Studio because Walt was so clearly the focus of attention.

It was in this studio that Disney produced the last of the Alice Comedies, all of the Oswald the Lucky Rabbit shorts, and the early Mickey Mouse and Silly Symphony shorts. In addition, this was where work was done on *Snow White and the Seven Dwarfs*, *Pinocchio*, *Bambi*, and some of *Fantasia*.

The next door neighbors in 1926 were a gas station and an organ factory. Although the area had been designed as a residential neighborhood, at the time the Disney Studio operated there, it was generally a quiet, empty location with little activity and lots of unattended landscaping.

In 1926, Walt and Roy filed a permit for a three-room artist studio that included two small offices for themselves, a camera room, and a large partitioned work area for the animators and ink and paint staff.

On the roof was an iconic neon sign identifying the studio as the home of Mickey Mouse.

During the next four years, the original studio building underwent several renovations and additions until a two-story building called "Animator's Building No. 1" and a sound stage were added in 1931. The studio had filled out the space between the organ factory in 1929 and by 1930 also extended out to the street.

Walt and Roy purchased additional plots of land surrounding the studio and built the "Animator's Building No. 2" (known as the Shorts Building) in 1934, Ink-and-Paint and Annex buildings in 1935, and a "Features Building" in 1937.

Over the 14 years the studio operated, several other smaller buildings were constructed on the property including a warehouse, film vault, sound stage monitor room, camera room, and of course, the famous garage that housed Mickey Mouse's car. The studio even rented nearby bungalows for the story staff, as well as for Charlotte Clark and her crew, who were making stuffed Mickey Mouse dolls.

It was apparent that the studio had become a hodgepodge of inefficient buildings and further expansion in the area would be impossible.

In June 1938, Walt and Roy paid a $10,000 deposit for a 51-acre plot of land in Burbank for the construction of a new studio to open in 1940.

In 1941, half of the Hyperion studio property was sold to Thomas Curtis Optical Laboratories for $75,000. The Weldon T. Thomas Company, a vitamin manufacturer, purchased the remaining property, basically the soundstage, for its business.

The original studio building along with the adjunct buildings were completely destroyed by a bulldozer in August 1966. A Mayfair Market grocery store was built (later replaced by Gelson's Market). There is a vintage photo of the Hyperion studio posted inside of Gelson's today.

A duplex from the former Hyperion studio property still stands in its original location and is a private residence. Before leaving the space in 1939, two Hyperion buildings were moved and combined to create the Shorts Building on the Burbank Lot.

The Personnel Building that housed the Burbank studio store and employee center was likewise moved from Hyperion to Burbank.

The Publicity and Comic Strips Building, which was actually a small wood frame bungalow, was moved to Burbank where it served as the studio mail room for many years, and was moved again and renovated as two conference and meeting spaces, known as the Hyperion Bungalow and the Silver Lake Room (Silver Lake is the neighborhood where the Hyperion studio was located).

The Hyperion location was declared a historical-cultural monument in October 6, 1976 and marker No. 163 was installed. In 2005, the landmark was amended to include more area (2660-64 North Hyperion Avenue, 2646-64 North Griffith Park Boulevard, 3027-33 Angus Street) that among other things added the space for Disney's first animation school.

Walt Disney's Burbank Studio
500 South Buena Vista Drive, Burbank, CA 91505

This is the corporate office for the Disney Company today. Some of the Disney studio staff moved in as early as December 1939, but the move from Hyperion was complete by May 1940.

While the studio's address is officially 500 S. Buena Vista, Walt's formal and working offices (3H; 3rd Floor of the Animation Building) were actually nearer to the corner of Buena Vista & Alameda. Mail for Walt Disney was most often delivered to his studio offices c/o 2400 West Alameda.

Walt Disney's biggest concern with his new studio was that each building should be designed to facilitate optimum production.

The Animation Building was built to house the Story Department, directors, producers, background artists, layout artists, in betweeners (animators who provide drawings between key poses, so as to provide a flow of movement), and similar personnel. Walt hoped having all these people together, even if on different floors, would help them to coordinate better with each other.

Roy Disney's office, along with the accounting and business department, had their own first floor wing. The animators, clean-up artists and inbetweeners were also on the first floor.

The directors had suites on the second floor, including a secretary and a separate room with a Moviola (a device that film editors used to view film while editing). Layout men, background artists and the music room were also on the second floor.

The third floor was Walt's suite of offices, as well as rooms for the Story Department and the music composers.

The Ink and Paint Building was built across from the Animation Building with an underground, all-weather tunnel to connect the two so that artwork could be transported without fear of wind, rain, or extreme heat.

Walt was adamant that he didn't want an "institutional" look very common in other studios, especially for the main Animation Building that was to be 250 feet long.

Frank Crowhurst was in charge of construction in 1940, supervising the architects and engineers. The primary designer was Kem Weber.

For the soundstages:

- Stage One was built in 1940 and the live-action sequences for *Fantasia* were the first scenes to be shot on that stage.

- Stage Two was built in 1949 in cooperation with film producer and actor Jack Webb, who used the stage to film his *Dragnet* television series.

- Stage Three was built in 1953 for *20,000 Leagues Under the Sea* and contained a large tank for underwater and special effects filming.

- Stage Four was built in 1958 and first used for filming *Darby O'Gill and the Little People*.

Some of Disneyland was built on these sound stages and trucked to the Anaheim location.

The studio water tower once held 150,000 gallons of water and was built at a cost of $300,000. Most studio water towers like the one at Warner Bros. have four support legs, but Roy O. Disney specifically insisted on six for this 135-foot tall structure.

Massive changes to the studio were made by Michael Eisner when he was CEO.

The fabled backlot was demolished and repurposed with office buildings. The feature animation staff relocated in 1985 to warehouses and trailers in the Glendale area. The Animation Building was converted into business offices in 1986. The animation staff returned in 1995 to a new building on Riverside Drive across the street from the studio, with an entrance designed to look like Mickey Mouse's hat in the Sorcerer's Apprentice segment of *Fantasia*, and dedicated to Roy E. Disney.

There is also a Roy O. Disney building that used to house the Disney Archives, but currently serves as offices for Disney Legal.

Michael Graves created an imposing Team Disney building with sculptures of the seven dwarfs supporting the roof in 1990. On January 23, 2006, it was rededicated as Team Disney—The Michael D. Eisner Building. The Frank G. Wells building opened July 1998 and, among other departments, now houses the Disney Archives.

Walt's Hollywood Theaters

Walt loved going to the movies, not only for inspiration and enjoyment, but because he was interested in the audience's reactions. He usually found a seat in the back of the theater so that he could observe.

In the *The Hand Behind the Mouse*, Leslie Iwerks and John Kenworthy wrote:

> [Walt Disney and Ub Iwerks] spent their spare time [in Kansas City, Missouri] haunting movie theaters...They studied any available movie—live-action or animation—that they could afford to see. Their schools were the theaters of Kansas City."

Walt's wife, Lillian, in an interview from *McCall's* magazine February 1953, recalled:

> The first time I remember Walt ever seeing one of his cartoon shorts in a theater was just before we were married [in 1925]. My sister and I were visiting a friend that night, so Walt decided to go to the movies. A cartoon short by a competitor was advertised outside, but suddenly, as he sat in the darkened theater, his own picture came on.

> Walt was so excited he rushed down to the manager's office. The manager, misunderstanding, began to apologize for not showing the advertised film. Walt hurried over to my sister's house to break his exciting news, but we weren't home yet. Then he tried to find Roy, but he was out, too. Finally, he went home alone. Every time we pass a theater where one of his films is advertised on the marquee I can't help but think of that night.

Besides the personal enjoyment Walt got from going to see movies, he often previewed his cartoon shorts at local movie theaters. The Alice Comedy *Alice's Balloon Race* was sneak previewed at the now demolished Bard's Glen Theater on Colorado Street in Glendale on December 11, 1925, to judge the audience's reaction. It officially was released to the general public on February 15, 1926.

Here are four other movie theaters that had prominent connections to Walt. Two of them are gone, but two are still showing films.

Alex Theater
216 North Brand Boulevard, Glendale, CA 91203

This movie theater opened in 1925 as the Alexander Theatre, which it was called until 1939. In 1940, the newly designed exterior had Alex on the marquee as the new official name. The theater was named after Alexander Langley, the son of C.L. Langley, owner of the West Coast chain that included the Raymond Theater in Pasadena.

From the 1920s through the 1950s, it was used as a sneak preview house for major Hollywood films, as well as its regular schedule of films. Singer-actor Bing Crosby nervously paced the lobby carpet during a preview screening of *Going My Way* in 1944 as he worried whether the movie-going public would accept him as a movie priest.

A teenager named Marion Morrison worked at a nearby soda fountain and his pal Bob McCaskey ushered at the Alex, where he let in the future John Wayne for free, in exchange for Morrison's "on-the-house" sodas.

Walt Disney, in the 1930s, often previewed his animated shorts at the theater a few miles from his Hyperion studio to judge audience reaction. He and his animators would sit in the back and then go out to the lobby or outside the theater for Walt to give his evaluation. It could be an anxiety-inducing experience as Walt would concentrate on what needed to be improved.

In their highly recommended (but difficult to obtain) book *Silly Symphonies: A Companion to the Classic Cartoon Series*, Russell Merritt and J.B. Kaufman were able to document that Disney had preview screenings at the Alex Theatre at least as early as 1931 with the Silly Symphonies series, including *Egyptian Melodies* and *The Clock Store*.

Disney Legend and veteran animator Ward Kimball stated:

> We always previewed our pictures in Glendale at the Alexander, and they let us know when they'd run *The Wise Little Hen* or *Orphans' Benefit* and we'd all go out. We had passes and we would sit in the audience and listen, and Walt would walk outside and have an impromptu discussion.

Clarence Nash, who supplied the voice of Donald Duck remembered that preview showing of *Orphans' Benefit*:

> We drove over to the Alexander Theatre, here in Glendale, for the preview. I was more nervous about that picture than I was about *The Wise Little Hen*. I was with a group of Disney people, and my wife was with me, too. I got a big kick out of it and completely forgot that I had anything to do with it.

In his book *Walt Disney and Other Assorted Characters*, animation director Jack Kinney confirmed that the location was used for previewing the animated shorts:

When a picture was finished, it was usually previewed at the Alexander Theatre in Glendale to get audience reaction. After the show, the boys and girls would gather in the lobby and discuss the various scenes with Walt.

Walt's final animated short directorial effort *The Golden Touch* (1935) was previewed at the Alexander Theater. The audience reaction was so weak that Walt didn't hold a discussion session afterward.

In 1992, the Glendale Redevelopment Agency purchased the Alex Theatre to serve as the centerpiece of Glendale's revitalized Brand Boulevard and embarked on a $6.5 million rehabilitation of the facility. It is now home to resident companies such as the Alex Film Society, Glendale Pops, Glendale Youth Orchestra, Los Angeles Ballet, and Los Angeles Chamber Orchestra, and hosts music, dance, theatre, comedy, film, and special events each season.

The Alex Theatre was added to the National Register of Historic Places in 1996.

Pasadena Crown Theater
129 North Raymond Avenue, Pasadena, CA 91103

The Raymond Theater opened in 1921 as a vaudeville and movie theater. In 1948, it was renamed the Crown Theater and became the premier theater in the city from 1948 to 1974, rivaling Grauman's Chinese Theater.

Like the Alex, this was a theater where Walt would preview some of his animated shorts to get the reaction from the audience.

Walt Disney's first True-Life Adventures documentary short *Seal Island* (1948) did not appeal to RKO-Radio Pictures who distributed Disney films to movie theaters at the time. They balked at releasing it. They felt audiences would not sit still for a nature film.

Walt asked a friend who ran Pasadena's Crown Theater to show the film for one week in December 1948 so that this nature film would qualify for consideration for an Academy Award nomination.

Though it was 27-minutes long (much longer than the usual short subject), *Seal Island* won that year's Best Documentary Oscar. The very next day, Walt took that Academy Award to his brother Roy Disney's office and said: "Here, Roy. Take this over to RKO and bang them over the head with it."

In 2004, despite a lengthy four-year lauded battle by the Friends of the Raymond Theater to prevent it from being demolished, developer Gene Buchanan, who owned the property, drove a bulldozer through the sidewall of the theatre and demolished anything possible to make sure the theatre could not be saved.

Today, what remains of the existing building is a mixture of apartments and retail use.

Carthay Circle Theater
6316 San Vicente Boulevard, Los Angeles, CA 90048

The Carthay Circle Theater was a movie palace designed in Spanish Mission Revival style by architect Dwight Gibbs. The 1,500 seat theater opened in 1926 in the mid-Wilshire district. It had a circular auditorium and the interior was adorned with forty foot tall murals of historical events.

Along with Grauman's Chinese Theater, the Carthay Circle hosted more big West Coast movie premieres than any other Hollywood theater. *Gone with the Wind* had its Hollywood premiere there in 1939.

In 1929, Walt decided to produce a new animated series, the Silly Symphonies, and the first installment was *The Skeleton Dance*. However, Walt's distributor didn't want to release the cartoon, but wanted more Mickey Mouse cartoons, instead.

Walt found a salesman he knew at a local pool hall, gave him a copy of the cartoon, and convinced him to contact Fred Miller, the owner of the Carthay Circle Theater. Miller liked the cartoon and booked it into his theater in August 1929, where it was a huge hit and gave Walt plenty of positive reviews to convince his distributor to book the film in other theaters.

It was this success that convinced Miller to take a chance on the first feature-length animated cartoon, *Snow White and the Seven Dwarfs*. It premiered with much hoopla at his Carthay Circle Theater in December 1937 with the nearby street median decked out as the dwarfs' cottage and mine.

The film had already been booked, sight unseen, as the Christmas attraction at Radio City Music Hall in New York, but Walt wanted a Hollywood premiere for his peers to help demonstrate that his work in animation was comparable to their work done in live-action films.

The Carthay Circle Theater was only one of only a handful of theaters to be fitted with the full Fantasound equipment for the premiere of *Fantasia* in 1940. Twelve of the 14 theaters showing the film were legitimate stage theaters and not movie theaters.

The theater was demolished in 1969. Today, two low-rise office buildings and a city park (called Carthay Circle Park at the corner of San Vicente and Crescent Heights just south of the Miracle Mile, roughly between Wilshire and Olympic and between La Cienega and Fairfax) occupy its former site.

A number of factors caused the theater's demolition, including earthquake damage, losing money in the age of multiplex theaters, and the developers' desire to build a huge office tower in its place.

However, the façade of the theater lives on both in a small gift shop that opened in 1994 on Sunset Boulevard at Disney's Hollywood Studios theme park in Florida and in a building on Buena Vista Street that opened in 2012 at Disney California Adventure. Neither replica is the same size or has the same interior floor plan of the original theater.

Grauman's Chinese Theatre
6925 Hollywood Boulevard, Hollywood, CA 90028

The theater opened in May 1927 after 18 months of construction at a cost of more than $2 million. It was meant to resemble a huge red Chinese pagoda, and was the dream of entrepreneur Sid Grauman, who was responsible for other movie palaces including the Egyptian Theater just down the street.

According to the theater's official website:

> Authorization had to be obtained from the U.S. government to import temple bells, pagodas, stone Heaven Dogs, and other artifacts from China. Poet and film director Moon Quon came from China, and under his supervision Chinese artisans created many pieces of statuary in the work area that eventually became the Forecourt of the Stars. Most of these pieces still decorate the ornate interior of the theatre today.

Sid Grauman sold his share to William Fox's Fox Theatres chain in 1929, but remained as the theater's managing director until his death in 1950. On March 24, 1949, Grauman received a special Oscar because he was a "master showman who raised the standard of exhibition of motion pictures".

Walt was friendly with Grauman and had him caricatured in the Mickey Mouse cartoon short, *Mickey's Gala Premiere* (1933). Two years later, the first Silly Symphony in color got its premiere at Grauman's on July 18, 1932.

In *Walt Disney: An American Original*, Bob Thomas wrote:

> After the first few scenes had been completed [on the Silly Symphony *Flowers and Trees* released in 1932], Walt showed them to a friend, Rob Wagner, publisher of a literary magazine in Beverly Hills. Wagner was so impressed that he invited Sid Grauman, impresario of Grauman's Chinese Theatre, to see the film.

> The film lasted only a minute, but Grauman said he wanted 'Flowers and Trees' to open with his next attraction, [MGM's] *Strange Interlude*, starring Norma Shearer and Clark Gable. Walt worked his animators overtime to finish ahead of schedule, and Technicolor sped the processing.

> When *Flowers and Trees* appeared at the Chinese, in July 1932, it created the sensation that Walt had hoped for. No longer was the Silly Symphony the neglected half of the Disney product. *Flowers and Trees* got as many bookings as the hottest Mickey Mouse cartoon. Walt decreed that all future Symphonies would be in color.

In 1973, the theater was purchased by Ted Mann, owner of the Mann Theatres chain. From then until 2001 it was known as Mann's Chinese Theatre.

Of course, the theater is perhaps best known for its forecourt with its imprints of hands, feet, and signatures of movie stars in concrete.

In 1984, a costumed Donald Duck, along with help from his long time "voice" Clarence "Ducky" Nash, imprinted a square in the forecourt on May 21, 1984, as part of Donald's 50th birthday celebration. It was the 150th ceremony. Neither Mickey Mouse nor Walt Disney were ever afforded that honor.

Through the years, the original Chinese Theatre has not only shown movies, but it has had its image used in many films. *Singin' in the Rain*, *Blazing Saddles*, and Disney's *The Rocketeer* are among those in which it has appeared.

The theater was declared a historic-cultural landmark in 1968, is still used frequently for red carpet premieres, and draws nearly four million visitors a year.

For the opening of the Disney-MGM Studios in Florida in 1989, the Disney Imagineers constructed an exact replica of the famous theater at the end of Hollywood Boulevard. They used the original 1927 Meyer and Holler blueprints for reference and the façade was built to full scale, rather than the forced perspective often used in such huge structures at Disney theme parks.

Disney did not have permission to use the names of either Grauman's Chinese Theatre or Mann's Chinese Theatre (as it was called in 1989), so the building is just called The Chinese Theatre. The Disney version was built to resemble how the theater looked in the late 1930s and early 1940s. It is 96 feet tall.

The pagoda roof alone stands 45-feet tall and weighs 22 tons. It was built separately and then hoisted into place by a crane. In the Florida version, the ticket booth that was in front of the entrance was moved to the left to provide a better view for the guests and to facilitate guest flow.

The two display areas in the front of the Great Movie Ride attraction feature photos and memorabilia of two Disney red-carpet movie premieres that happened at Grauman's Chinese Theatre: *Mary Poppins*, on August 27, 1964, which Walt and his wife attended, and *The Jungle Book*, on October 18, 1967, less than a year after Walt's death. However, with the new sponsorship of the attraction by TCM, those displays are slated for removal.

Walt's Hollywood Hangouts

Walt Disney appeared frequently at many locations in the Los Angeles area. Here are a few of the places he was known to visit quite often.

Tam O'Shanter
2980 Los Feliz Boulevard, Los Angeles, CA 90039

Los Angeles' oldest restaurant that has remained in the same location under the same ownership and management since 1922 was located very close to Walt's Hyperion studio when it opened.

Since Disney's studio did not have a commissary, Walt and his staff often went to this restaurant for lunch with such frequency that it was jokingly referred to as the "studio commissary". Walt's favorite table was No. 31 near the fireplace. Actor John Wayne's favorite table was No. 15.

Walt often used the location for meetings, as well. In 1934, Walt had lunch with a former McKinley High School and Chicago Art Institute classmate at the Tam O'Shanter. That classmate was Bianca Majolie, and Walt was so impressed with her artistic talents and story skills that he hired her as his story department's first female employee.

The restaurant opened under the name Montgomery Inn in 1922. In 1925, it was renamed The Tam O'Shanter and was re-themed as a Scottish inn. It was one of the first themed restaurants with tartans and other artifacts decorating the interior.

Scottish writer Robert Burns introduced the character of Tam O'Shanter in a poem in 1790, and the term has since come to be associated with a voluminous Scottish-style wool hat worn by men.

The "Tam-O", the nickname for the restaurant, was co-owned by Lawrence L. Frank (later the founder of Lawry's Prime Rib Restaurants) and his father-in-law Walter Van de Kamp (famous for Van de Kamp bakeries).

It became a hang out for the stars of the silent film era, like Tom Mix, Fatty Arbuckle, and Mary Pickford, and would continue to be popular with Hollywood celebrities through the 1960s. It was known for its prime rib.

In 1968, the restaurant underwent a name change to The Great Scot, but found its way back to the original namesake in 1982 on its 60th anniversary.

On one wall is a colored sketch (drawn by Disney Legend John Hench, but signed by Walt thanking Frank) of Lawrence Frank in tam and tartan surrounded by Mickey and Minnie Mouse, Goofy, Donald, and Pluto also dressed in similar garb.

Another drawing, done after Walt's passing, is a color sketch of Richard Frank Sr. (Lawrence's son), at the head of a table of food. Seated on either side of the table are images of Mickey Mouse as he appeared over the decades.

Another wall has various photos of Disney Legends Marc Davis, Ward Kimball, Frank Thomas, and Ollie Johnson on one of their last visits together to the Tam-O.

The "Storybook Style" architecture of the restaurant was designed by Harry Oliver, a movie art director in the 1920s and 1930s, who used movie studio carpenters to help build it.

Oliver remained a friend of Walt's until the latter's death. Walt publicly stated that it was Oliver who coined the word "litter bug". Walt had the artists at his studio draw up a caricature of Oliver riding on a donkey for a trashbag as part of the "Keep the Desert Beautiful" campaign.

While there have been some cosmetic and structural changes over the years, the integrity of the original concept and design theme of the Tam O'Shanter is much the same as it was when Walt dined there.

Hollywood Brown Derby
1628 North Vine Street, Hollywood, CA 90028

There were four Brown Derby restaurants in the Los Angeles area (Hollywood, Beverly Hills, Los Feliz, and Wilshire Boulevard), but only the one on Wilshire Boulevard was in the shape of a derby hat.

The one most frequented by Walt was on Vine Street, a half block south of Hollywood Boulevard, that opened in 1929 and was operated by Bob Cobb. It was in the center of broadcasting studios, theaters, and movie studios, so it became a popular location for celebrities and for making deals.

The architect was Carl Jules Weyl who later became a Warner Bros art director after the Depression. He designed the iconic Rick's Café in the classic film *Casablanca* where Rick (Humphrey Bogart) had an office above the restaurant (just like one that Weyl designed for owner Bob Cobb above the Hollywood Brown Derby).

It was at this Brown Derby that both the Cobb salad and the infamous grapefruit cake were created. Movie stars often received fan mail addressed simply to their name c/o The Brown Derby, Hollywood and Vine.

Besides dining there often with his wife, Walt and Cobb were friends who also shared a love of baseball. They served together on the board of directors of the Pacific Coast League's Hollywood Stars.

Diane Disney Miller recalled:

> Mother and Dad had a box at Gilmore Field [located near Fairfax and Third Street], which was the home of the Hollywood Stars of the Pacific Coast League.
>
> This was during the 1940s and 1950s. The box was right behind the Stars' dugout, between first base and home plate. They attended most of the home games, and when I was in my early teens, I went with them to night games during the week and double headers on Sunday afternoons.
>
> When Walter O'Malley moved the Brooklyn Dodgers to Los Angeles, the Hollywood Stars ceased to exist.

Walt and Cobb were on the advisory board of Gene Autry's California Angels baseball team, as well.

Disney Miller continued:

> When Gene Autry formed the Los Angeles Angels, he offered to let Dad and Cobb purchase some stock and credited them as "advisers". When I noticed this on the game program once, I was really impressed. Dad acknowledged it, and said, with mock ruefulness, "But he hasn't asked me for any advice yet."

In the Disney 1947 film *Fun and Fancy Free*, at the end of the "Mickey and the Beanstalk" segment, Willie the Giant stomps through Hollywood looking for Mickey Mouse. The giant sees the Wilshire Brown Derby restaurant shaped like a derby all lit up in the evening and picks it up and puts it on his head.

The restaurant was closed in 1985 from fire and earthquake damage and demolished in 1994.

The Hollywood Brown Derby began its licensing program in 1987 with an agreement with the Disney Company for a re-creation of the restaurant at Disney's Hollywood Studios (then the Disney-MGM Studios) at Walt DIsney World.

In 1990, Disney entered into three additional agreements for Euro-Disney, Tokyo Disney, and Disneyland in Anaheim, California.

Actor Charlton Heston's star on the Hollywood Walk of Fame is directly in front of the original entrance to the restaurant. There is also historical marker #22 nearby telling the story of the famous eatery.

Patys
1001 Riverside Drive, Toluca Lake, CA 91602

Opened in 1960, Patys was known for diner favorites and home-baked goods. It was a popular location of many locals including stars like Bob Hope, Johnny Carson, and James Garner. Today, it is a well-known "retro diner" and many current celebrities like Steve Carell, Zac Efron, and Hilary Duff go there to eat.

Disney Legend and composer Buddy Baker once owned the restaurant and remembered that Walt would linger with a cup of coffee at the counter to watch the baker make mini loaves of bread. According to Baker, Walt found this so fascinating that he considered including the process at Disneyland where guests could watch through a plate glass window and then purchase the results, something that was incorporated into the Main Street Candy Palace.

Biff's
Burbank, CA

"[Walt] liked to eat at Biff's [a little coffee house on a nearby corner near the Burbank studio]. He felt they did their potatoes 'right' by pan-frying them," said Diane Disney Miller, who also revealed that the Disney family cook, Thelma, went to check and the potatoes were just hash browned.

Restaurant chain owner W.W. "Tiny" Naylor (the nickname "Tiny" came from him being 6'4" and weighing 320 pounds) started Biff's (his son's name) in 1948. He also started the Tiny Naylor "Googie-style" coffee shops in 1957. There were Biff's locations in Burbank, Santa Monica, Reseda, Northridge, Encino, Long Beach, Hollywood, and elsewhere.

The chain offered typical diner food like hamburgers, hot roast beef sandwich, tuna melt, steak sandwich, fish and chips, and chili.

The cover of the Biff's menu was drawn by famous cartoonist George McManus (*Bringing Up Father*) and featured an overweight cook stepping forward carrying high above his head with his left hand a steaming pot with a smiling duck's head and neck sticking out.

The North Hollywood location at Magnolia Boulevard and Lankershim Boulevard was often used in the television series *Adam-12* (for example, season 1, episodes 6 and 25) as a hangout for the police officers, since it was close to the studio where the series was being filmed. The interior and exterior shots look pretty much like the Biff's that Walt would have visited.

St. Joseph's Hospital
501 S. Buena Vista Street, Burbank, CA 91505

The hospital was founded in 1943 by the Catholic nuns the Sisters of Providence. Walt was a strong supporter of the hospital. It was located directly across the street from the Disney Studio and is still there today.

On November 30, 1966, Walt collapsed and was admitted to the hospital under the pseudonym "John Smith" so as not to alert the press. Walt died at 9:35 am on December 15, 1966, in Room 529, at the age of 65.

The cause of death, according to Walt's official death certificate, was cardiac arrest due to bronchogenic ca(rcinoma), or more commonly, lung cancer. The certificate also shows that Walt was cremated on December 17 at Forest Lawn Memorial Park Glendale.

Walt's older brother Roy collapsed in December 1971 and was taken to St. Joseph's Hospital. He was in room 421. He had suffered a massive brain hemorrhage and was put on life support. His wife, Edna, consented to have him removed from the equipment and he died at the age of 78 on December 20, 1971.

Roy's grandson, Roy Patrick Disney, had taken a bad fall from a tree a couple of days earlier and suffered severe head injuries. He was in room 321, just one floor below his grandfather. It was touch and go for a while, but Roy Patrick recovered.

Edna Disney died at St. Joseph's on December 19,1984.

Other celebrities, like animation legend Tex Avery and voice of Donald Duck Clarence Nash, died at St. Joseph's, as well.

Forest Lawn Memorial Park Glendale
1712 S. Glendale Avenue, Glendale, CA 91205

Hubert Eaton took over the management of the cemetery in 1917. He envisioned Forest Lawn to be not a cemetery, but a memorial park "devoid of misshapen monuments and other signs of earthly death, but filled with towering trees, sweeping lawns, splashing fountains, beautiful statuary, and...memorial architecture". The grounds are adorned with 1,500 statues, some of which are replicas of famous works of art.

Forest Lawn is a popular tourist attraction, with nearly one million visitors each year, because of its beauty and serenity as well as the many celebrities that are interred there. Prior to the opening of Disneyland, it was the most popular tourist attraction in Los Angeles.

Walt's mother and father loved spending a day just walking the grounds. Forest Lawn Cemetery also presented various religious shows. It was later discovered that their visitor patterns paralleled, on a smaller scale, those of Disneyland

A small private memorial service was held at 5 p.m. on December 16, 1966, at the Little Church of the Flowers for Walt Disney. Only his wife, Lillian; his daughters, Diane and Sharon (and their husbands); his brother, Roy (and his wife, Edna); and his nephew, Roy Edward (and his wife Patty) attended. Walt's younger sister, Ruth, did not attend for fear that she would be hounded by the press if she left Oregon where she was living.

Walt had been insistent that he did not want a public funeral. The studio gave out no information other than the service was private and that the family requested, instead of flowers, that contributions be made to the California Institute of the Arts.

According to the official death certificate, Walt's body was cremated the following day, December 17.

For nearly a year after the cremation, Walt Disney's ashes remained un-interred. When Sharon's husband, Bob Brown, died less than a year later, in September 1967, Sharon made the arrangements for her father and her husband to be interred together so that neither would be alone. She and her older sister, Diane, chose a remote plot outside the Freedom Mausoleum.

A modest bronze rectangular tablet on a wall lists the name of Walter Elias Disney; his wife, Lillian; his son-in-law, Robert Brown; and a mention that daughter Sharon's ashes were "scattered in paradise".

To locate the site, drive through the entrance to a road called Cathedral Drive. Stay on the road to the eastern edge of the park where Cathedral Drive intersects with Freedom Way. At that intersection, turn right onto Freedom Way. On your left will be trees, fountains, and statues. This area is called Freedom Court.

At the far end of Freedom Court is a large mausoleum. Pull over and park on the right-hand side of the street. There should be a "33" painted on the curb opposite your car, indicating 33 Freedom Way. Standing at the base of the steps leading to the main entrance of the Freedom Mausoleum, turn to your left and walk to the far edge of the steps.

There is a small, private, low-gated courtyard garden near the brick wall. Inside this area guarded by a hedge of orange olivias, red azaleas, and a holly tree. There is a plaque on the wall. You will see a small statue of Hans Christian Anderson's Little Mermaid sitting on a rock.

Walt Disney was not cryogenically frozen.

Walt's father, Elias, is at the Great Mausoleum, Memorial Terrace, Sanctuary of Truth, Crypt 5499, right next to his wife, Walt's mother Flora.

Walt's brother Raymond is at the Freedom Mausoleum, Sanctuary of Prayer, Crypt 20400.

Forest Lawn Memorial Park Hollywood Hills

6300 Forest Lawn Drive, Los Angeles, CA 90068

Roy Disney was not cremated but buried at the Forest Lawn that overlooks the Disney Studio. His grave is across the park, near the main gate in the Sheltering Hills section.

Drive southeast a short distance until you see a white statue of two small children (on the right side of the road). Starting from that same white statue of the two children, walk up (west) seven rows, then go left about 10 spaces to grave No. 125. (Or use the directions: seven rows up from Memorial Drive and nine markers in from Evergreen Drive.) Roy's grave marker reads: "A great and humble man. He left the world a better place."

His wife, Edna, is buried next to him. Her marker reads: "Loving wife, devoted mother, and a gallant lady."

Hollywood Walk of Fame

The first stars on the Hollywood Walk of Fame were installed in spring 1956 and now include almost 2,500 stars embedded in the 15 blocks of Hollywood Boulevard and three blocks of Vine Street in Hollywood.

While several stars are dedicated to Disney characters, three stars acknowledge the Disney brothers:

- Walt Disney—7021 Hollywood Boulevard (February 8, 1960; star for Movies)
- Walt Disney—6747 Hollywood Boulevard (February 8, 1960; star for Television)
- Roy O. Disney—6833 Hollywood Boulevard (July 24, 1998; star for Movies)

Walt's Bust

5220 Lankershim Boulevard, North Hollywood, CA 91601

Walt Disney was inducted into the Television Academy Hall of Fame in 1986. His widow, Lillian, accepted the award. A bust of Walt Disney sculpted by Disney Legend Blaine Gibson is in the courtyard. A duplicate is in the re-creation of the courtyard at Disney'sHollywood Studios near the Hyperion Theater.

Griffith Park Merry-Go-Round and Walt's Barn

4730 Crystal Springs Drive, Los Angeles, CA 90027

Griffith Park is a large municipal park located in the Los Feliz neighborhood of Los Angeles, where Walt lived for most of his life. In December 1896, Colonel Griffith donated five square acres of his Rancho Los Feliz estate to Los Angeles to be used as "a place of recreation" for "the plain people".

During Walt's lifetime, there were several attractions located on the park grounds, including an observatory, a zoo, a theater, and pony rides, as well as a special section dedicated on December 1952 called Travel Town, an outdoor museum to preserve and celebrate southern California railroads with actual vehicles that people could explore.

More important, there was the Griffith Park Merry-Go-Round. Built by the Spillman Engineering Company in 1926 and brought to Griffith Park in 1937, the merry-go-round has 68 horses with finely carved, jewel-encrusted bridles, detailed draped blankets, and sunflower and lion's head decorations.

A Stinson 165 Military Band Organ, reputed to be the largest band organ accompanying a carousel on the West Coast, plays over 1,500 selections of marches and waltz music.

Walt would take his two young daughters on the weekend to ride the carousel. Supposedly, while sitting on a nearby bench, he decided to build a place where parents and children could have fun together.

Adjacent to the Travel Town Museum, at 5202 Zoo Drive, is the section operated by the Los Angeles Live Steamers Railroad Museum. Part of the facility includes the barn workshop that was in the backyard of Walt's Holmby Hills home on Carolwood Drive.

Walt the Bibliophile

"There is more treasure in books than in all the pirates' loot on Treasure Island and at the bottom of the Spanish Main...and best of all, you can enjoy these riches every day of your life," stated Walt Disney for *Wisdom* magazine (Volume 32, 1959).

Though he only had one year of high school education, Walt loved books. The books that he read as well as the books he had in both his home and office libraries waiting to be read helped define who Walt Disney was.

Walt didn't like the holidays because he couldn't work, Lillian Disney noted, and even at Christmas "when he got through with the festivities, he went to his room and read," according to a December 27, 1937, *Time* magazine article, "Mouse and Man".

Walt was primarily reading scripts that he brought home each night in a battered brown briefcase that is now on view in his "working" office re-creation at the One Man's Dream attraction at Disney's Hollywood Studios at Walt Disney World. Walt brought home stacks of scripts to read in the living room or on the porch. Because it sometimes pained him to sit up straight, he would often put them in his lap and leaned over to read them.

Diane Disney Miller told me:

> There is a wonderful series of photos, taken by Earl Theissen in our Los Feliz Hills home, of Sharon and me leaning over dad in his favorite chair as he reads to us. The Siamese cat is on his lap, the poodle DeeDee on the floor in front of us [she's only visible in one], and he is reading to us from...a script!
>
> From the time I can remember, newspapers and scripts were definitely the bulk of his reading material. I do recall Dad reading to us, both of us, curled up on his lap but Mother was the one who, when I was quite young, sat by my bed almost every evening and read to me from various storybooks. This had to have been before I could read myself but, then, kids do like to be read to, even when they can read very well themselves.

Walt was curious about everything and although he was always a very visual person and had to see something rather than just read about it, Walt did quite a bit of reading during his life.

Disney Miller continued:

I do believe that dad was a life-long reader, even though he was a rather dismal performer in school for the reasons we all know. His love of story, of history, and his sense of what makes a story work best would have come from the experience of reading.

Over the years, there were many publicity photos of Walt with books. Often, he is holding an open book related to his latest film and apparently sharing the contents with his daughters or the young stars of the film or even the monkeys during the filming of *Monkeys Go Home!* (1967) These books sometimes look larger than real books, but they sure made a great picture.

As a young boy, Walt read all the works of Mark Twain. Hannibal, Missouri, where Samuel Clemens grew up, was only about sixty miles away from where Walt grew up. Things that Twain wrote about were pretty similar to things Walt had personally experienced.

Walt said his mother, Flora, read the Disney children to sleep by candle-light, but never mentioned what books she read. Walt's father, Elias, who was very religious, told his sons that if they had to read, to stick to the Bible.

At Benton Elementary School, Walt's first period was English and he said he especially liked the stories by Sir Walter Scott (*Ivanhoe*), Charles Dickens (*A Tale of Two Cities*), and Robert Louis Stevenson (*Treasure Island*). Walt also liked Shakespeare, but primarily the parts where the characters fought great battles and duels. "You'll be a poorer person all your life if you don't know some of the great stories and the great poems," Walt said in 1959 in *Wisdom* magazine.

All of these selections were in the McGuffey Eclectic Readers, a popular series of school books edited by William Holmes McGuffey. McGuffey edited the first four Readers in 1836-1837 and the final two were created by his brother, Alexander, in the 1840s. The series consisted of stories, poems, essays, and speeches. The Advanced Readers contained excerpts from the works of great writers like John Milton, Shakespeare, Poe, Sir Walter Scott, and Louisa Mae Alcott.

Each volume was progressively more difficult. The first reader taught reading by using the phonics method. Later, Walt would be embarrassed to read in public because he moved his lips when he read, the result of being taught to "sound out" the word by this method.

Diane Disney miller recalled:

> He did sort of "lip read". I've thought about that, and it could have been only on dialogue passages. I find myself doing that sometimes, to help get a better sense of the words the way they might have been said. We had a *Treasury of the McGuffey Readers* in our home ever since I can remember. I used to read through it.

The second reader helped understand the meaning of sentences with vivid stories. The third reader taught the definition of words (about fifth-and sixth-grade level), and the fourth began with punctuation and articulation and then presented 90 selections written by Daniel DeFoe, Louisa Mae Alcott, and others. The fifth and sixth readers featured selections from the Bible and classic American and British authors.

Henry Ford said that McGuffey Readers were one of his most important childhood influences. Revised versions of McGuffey Readers were used in schools through 1960 and are sometimes still used today by parents for home schooling purposes.

Young Walt was engrossed by the Horatio Alger's stories of impoverished boys who rose to success due to hard work, determination, courage, and honesty. He also devoured the adventures of Tom Swift, a series of novels about a young boy who uses science, inventions, and technology to save the day. Another of Walt's favorites was a now-obscure adventure series from the pulp magazines featuring Jimmie Dale, the infamous "Gray Seal", created by Frank Lucius Packard, a Canadian novelist born in Montreal, Quebec.

Dale appeared in serial installments in *People's Magazine*, *Short Stories Magazine*, and *Detective Fiction Magazine* before they were compiled into novels: *The Adventures of Jimmie Dale* (1917), *The Further Adventures of Jimmie Dale* (1919), *Jimmie Dale and the Phantom Clue* (1922), *Jimmie Dale and the Blue Envelope Murder* (1930), and *Jimmie Dale and the Missing Hour* (1935). A copy of *Jimmie Dale and the Blue Envelope Murder* appears prominently on Walt's desk at the Hyperion studio in a 1935 publicity photo.

Dale was an amateur private eye/safecracker in New York. His secret identity was the two-fisted Gray Seal, who could open even the most tightly guarded safes and who left his calling card, a gray diamond paper seal, but never stole anything.

In 1952, Walt bought the rights to the Jimmie Dale stories to develop into a television series. Several people at the Disney Studio remember Walt acting out the stories over the years.

To study animation, Walt borrowed the book *Animated Cartoons* (1925), by E.G. Lutz, from the Kansas City library and recommended it to others, including Hugh Harman and Rudy Ising, who worked for him on the Laugh-O-Gram animated cartoons.

Walt said:

> Everyone has been remarkably influenced by a book, or books. In my case it was a book on cartoon animation. I discovered it in the Kansas City Library at the time I was preparing to make motion-picture animation my life's work. The book told me all I needed to know as a beginner—all about the arts and the mechanics of making drawings that move on the theater screen.
>
> From the basic information I could go on to develop my own way of movie storytelling. Finding that book was one of the most important

and useful events in my life. It happened at just the right time. The right time for reading a story or an article or a book is important. By trying too hard to read a book that, for our age and understanding, is beyond us, we may tire of it. Then, even after, we'll avoid it and deny ourselves the delights it holds.

On a 1935 trip through England, France, Switzerland, Italy, and Holland, Walt brought back with him children's books with illustrations of little people, bees, and small insects. In a memo, Walt wrote: "This quaint atmosphere fascinates me and I was trying to think of how we could build some little story that would incorporate all of these cute little characters."

Nearly 700 books became the foundation of the Disney library that began that same year under the supervision of Helen Ludwig Hennesy. Even then, Walt realized the importance of having books available for his artists for reference and inspiration:

> It has always been my hope that our fairy tale films will result in a desire of viewers to read again the fine old original tales and enchanting myths on the home bookshelf or school library. Our motion-picture productions are designed to augment them, not to supplant them.

What was the Walt's personal family library at home like? Diane Disney Miller told me:

> We had a set of the Harvard Classics in the bookshelf of our library, which, together with the guest room behind it became our projection room when he began *So Dear To My Heart* and *Song of the South*. The bookshelves stayed. That's where I found *David Copperfield* and *Vanity Fair*. Loved Dickens, but was not intrigued with the story of *Becky Sharp* and didn't finish it. *Ben Hur* might have been in that set, because I read it at that time, but nothing else that I can recall in that series. I'm not certain if Dad did. We had a beautiful set of other classics.

> They're in our San Francisco bookshelves [at the Walt Disney Family Museum] now. Oscar Wilde's *Salome*, illustrated by Aubrey Beardsley; Voltaire's *Candide*; *The Rubiyat of Omar Khayam*, AE Housman's *Shropshire Lad*. Can't recall the other titles. He also had books that people he knew had written and inscribed to him.

> This was the time of the Writer's Club in Hollywood that Bill Cottrell told me about. Dad encouraged his writers to attend lectures, and often invited the lecturers to the studio for the benefit of his writers and animators. H.G. Wells, Rupert Hughes, Aldous Huxley. We have these books, too. We did have *Encyclopedia Brittanica*, too, of course.

Walt's home library was an eclectic mix of fiction, non-fiction, fairy tales and children's stories alongside books about history and nature and animals and other topics. His home library consisted of over a hundred different titles. Walt would find books anywhere. According to publicity for the film, he picked up a copy of *Mary Poppins* on the bedside table of his daughter, Diane, and that got him to thinking about adapting that story. Disney Miller realled:

> The *Mary Poppins* book, now an artifact in our exhibit, was sent to him by the publisher, who inscribed it: "Dear Mr. Disney... This is not Mickey, but we think you'll like our Mary." It had been in our home years before I could read, and I'm certain that he'd read it before he picked it up from my bedside table.

On the Disney family cruise through British Columbia in 1966, the last trip with the entire family, Ron Miller (husband of Diane) and Bob Brown (husband of Sharon) would go salmon fishing while Walt spent time on deck with books about city planning or universities as he developed his plans for Epcot. Walt was especially fond of *The Heart of Our Cities* by Victor Gruen and *Garden Cities of To-morrow* by Ebenezer Howard (re-issued in 1965).

Roy Disney enjoyed reading American history and amassed a large collection of works about Thomas Jefferson.

Every week on the Disney television program, Walt would casually go over to these massive book shelves on the sides of his formal office and, just as casually, immediately pull out just the right book that related to the episode that week.

What a marvelous, magical bookshelf that always seemed to have just the right book each week, from so many different categories, in just the right location for Walt to pull out!

Of course, that action was expertly staged for the needs of the episode.

Some of those books that Walt pulled out never actually existed. In the November 30, 1955, episode entitled "The Story of the Animated Drawing", Walt pulled out a copy of the book *The Art of Animation* and showed some samples of the amazing pages of early animation history. That book had not yet been written, and the final 1958 version by Bob Thomas bore little resemblance to that terrific book Walt had shown.

Disney Archivist Dave Smith personally catalogued all the real books that were on the shelves of Walt's formal office at the Disney Studio on the day Walt died. The list is 24 pages long, single-spaced, and contains more than 600 books. Smith has pointed out that the function of the bookcase in real life was somewhat more ornamental. Official visitors would come in and be overwhelmed with the wall of books.

Smith told me:

> Most of the books were put there, often without Walt's knowledge, by his secretaries. People were constantly sending autographed copies of their books or sending books hoping Walt would consider them for some future film project. I assume they might have been easier

to locate that way if someone came in to visit Walt in his office and wanted to know about the book they sent.

The vast majority of the list was indeed books autographed to Walt and there was no rhyme or reason to how the books were organized on the shelves. How do you explain a copy of *Les Aventures de Tintin, Objectif Lune* (Casterman 1953), but no other copies of the popular Tintin books or English translations? How do you explain that there is a copy of the official souvenir program for the Seattle World's Fair in 1962, but nothing from either the 1939 or 1964 New York World's Fairs?

Scattered on the shelves were copies of *We, the People, the Story of the United States Capitol* (U.S. Capitol Historical Society 1963); *The United States Polo Association Yearbook 1950*; *Pictorial Forest Lawn 1953*; a book in the Arabic language containing a mention of Disneyland; and *Celebrity Recipes* by Helen Dunn.

Autographed copies of the following books poked out in the stacks: *You Can Live Longer Than You Think* by Daniel Munro 1964, *Smoking in Your Life* by Alton Ochsner 1965, *Introduction to Tomorrow* by Robert Abernathy 1966, *Mardi Gras* by Robert Tallant, and *The Parables of Kahlil Gibran* by Annie Salem Otto in 1963.

There were books with obscure children's stories, non-fiction tomes on everything from ships to politics, and foreign editions. The sole animation-related book was an autographed copy of *L'esthetique du dessin anime* by Marie Therese Poncet from 1952.

Here is a brief overview of some of the books on Walt's shelves:

- The creator of Jimmie Dale, Frank L. Packard, autographed a copy of his 1925 book, *Running Special* to Walt that was on the shelf. Oddly, it was not a Jimmie Dale book.

- *I Love Her, That's Why!* by George Burns, about his love for his wife, comedian Gracie Allen, in a 1955 book published by Simon and Schuster and autographed to "Lillian and Walt".

- Pete Martin, who wrote Walt's biography with Diane Disney, gave Walt an autographed copy of his 1948 book, *Hollywood Without Make-up*. Roald Dahl, who worked briefly with the Disney Studio developing his story about gremlins, gave him an autographed copy of his 1948 book, *Some Time Never*.

- *L'Historie de Walt Disney*, the 1960 Hachette edition, and *A Historia de Walt Disney*, the 1960 Casa Editora Vecchi Ltda. edition, share space with the original *The Story of Walt Disney* by Diane Disney Miller (with Pete Martin), the Henry Holt 1956 hardcover edition. The copy in English is autographed by Diane.

- Jenny Disney Harcourt autographed a copy of Lewis Carroll's *Alice's Adventures in Wonderland* to her grandfather. On another shelf is *Alice's Adventures Under Ground* by Carroll, a facsimile of the original manuscript from University Microfilms Inc., 1964.

- Ruth Plumly Thompson autographed a 1934 copy of her *Speedy and Oz* book to Walt, which is not surprising since Thompson was continually pitching that Walt should produce her Oz stories. Also on Walt's bookshelf was a copy of *Who's Who in Oz* by Jack Snow in 1954, autographed by Thompson at the time Walt was considering making *The Rainbow Road to Oz* with the Mouseketeers.

- Roy Williams, the big "Mooseketeer" and Disney gag man, autographed to Walt a copy of his 1957 Bantam paperback, *The Secret World of Roy Williams*, a compilation of some of his one-panel gag cartoons. The book included this dedication: "To the man who has meant the most to me in faith and inspiration: Walt Disney whose patience and guidance through a lifetime of association are in the greatest degree imaginable responsible for the best of Roy Williams."

- Artist Mary Blair autographed a 1955 copy of *The Golden Book of Little Verses* by Miriam Clark Potter which was illustrated by Blair.

- O.B. Johnston had autographed a copy of *American Locomotives 1871–1881*, edited by Grahame Hardy in 1950. Johnston became head of Disney Merchandise after Kay Kamen died in a plane accident in 1949.

- Hardie Gramatky autographed a copy of his *Little Toot on the Thames* from 1964. For six years, until 1936, Gramatky had worked as an animator at the studio, and Disney did adapt an animated version of his original "Little Toot" story for the package film *Melody Time* (1948).

- Alfred Millote autographed a copy of his *The Story of a Platypus* (1959) and *The Story of a Hippopotamus* (1964). Millote was one of the True-Life Adventure photographers.

- Burl Ives autographed a copy of his 1962 book *The Wayfaring Stranger's Notebook*. Ives appeared in the Disney film *Summer Magic* (1963) around this time.

- Louis Rosenstein, "an old McKinley High School classmate", autographed a 1962 copy of *Jewels for a Crown* by Miriam Freund.

- Major Alexander de Seversky autographed a copy of his 1950 book, *Air Power: Key to Survival*, probably in gratitude for the 1943 Disney film *Victory Through Airpower*, based on another of his books.

- Walter Knott, who founded Knott's Berry Farm in Buena Park, California, and was highly conservative, autographed a copy of *Conscience of a Conservative*, the 1960 book by Barry Goldwater, to Walt, a staunch Goldwater supporter.

- There were a handful of other Republican-themed books on the shelves, although Walt also had *A Tribute to John F. Kennedy* by Pierre Sallinger from 1964.

- Walt had the *Britannica Book of the Year* volumes from 1954–1966 (but missing the volume for 1965, curiously). He also had the *International Picture Almanac* volumes from 1949–1964, *International Television Almanac* from 1958–1966, and the *Film Daily Yearbook* from 1957–1960.

- There were several foreign editions of the True-Life Adventures series, including two copies of *Lions d'Afrique* by Jean de'Esme (Librairie Payot, 1955), another copy of that book with the notation "special binding and printing for Walt Disney", two copies of *Perri* credited to Roy E. Disney (Centro Internazionale del Libro 1958 and Nouvelles Editions S.A., 1958), *Le Secrets de la Vie* by Julian Huxley (Nouvelles Editions, 1957), two copies of *Grand Prairie* by Louis Bromfield (Librairie Payot, 1955) with the notation "special binding and printing for Walt Disney", *Desert Vivant* by Marcel Ayme (Societe Francaise du Livre, 1954), *Desierto Vivente* (Central Peruna de Publicaciones), and *Deserto Che Vive* (Vellechi).

- Some books about the People and Places series were also on the shelves: *Siam* by Walt Disney (Bluchert Verlag, 1956), right next to *Im Tal Der Biber* by Walt Disney (Bluchert Verlag, 1963).

- *Walt Disney's Treasure Island* (Collins, 1950), one of the few motion-picture adaptations on the shelves, along with Lawrence Watkins' paperback adaptation of *Darby O'Gill and the Little People* (Dell, 1957).

- *Uncle Remus, His Songs and His Sayings* by Joel Chandler Harris (D. Appleton and Co., 1881) is on the same shelf as *Winnie the Pooh* and *The House at Pooh Corner* by A.A. Milne (both E.P. Dutton and Co., 1961) and *Just So Stories* by Rudyard Kipling (Doubleday Doran and Co., 1934). The Gordons autographed a 1963 copy of *Undercover Cat* that was the basis for the movie *That Darn Cat* (1965).

- Upton Sinclair autographed a 1936 copy of his *The Gnomobile* to Walt, as well as a 1962 copy of *The Autobiography of Upton Sinclair*. He also autographed a copy of *El Gnomomvil* from Ediciones Toray, S.A., 1964.

- Sterling North autographed a copy of his 1966 book *Raccoons are the Brightest People*, as well as *Rascal* from 1964.

- Pamela Travers autographed a copy of *Mary Poppins in the Park* (Peter Davies' 1958 edition of the book, which had originally been released in 1952), as well as *Mary Poppins from A to Z* (1962). She wrote: "To Walt Disney, hoping that your association with Mary Poppins will bring you joy & satisfaction & be as she herself has so often put it—a Pleasure and a Treat! With greetings from P. L. Travers, June 1961."

- T.H. White autographed a copy of *Verses* (Alderney, 1962), actually copy No. 26 out of 100 copies. Ward Greene autographed a 1953 copy of *The Lady and the Tramp*. Sheila Burnford autographed a copy of her 1960 book, *The Incredible Journey*.

- *The Magic Worlds of Walt Disney* by Robert de Roos (reprinted from *National Geographic*, August 1963, and "specially bound for Walt Disney") is on the shelves as well as an album labeled "To Walt Disney in grateful recognition of his participation in the Fourth of July celebration in Rebild National Park, Denmark, July 4, 1961"; two copies of the Beverly Hills B'nai B'rith "Testimonial Dinner to Mr. Walt Disney, Man of the Year, 1955"; an album labeled "A Resolution to Walt Disney" (Walt Disney Elementary School, Anaheim, on Disneyland's 10th anniversary, 1965); and an album labeled "A Walt Disney, cordial recuerdo de su admirador y amigo, F. Molina Campos".

- Four volumes of Miguel de Cervantes *Don Quijote de la Mancha* took up a lot of space.

- *Captain Danger* by Davis Crittenden is autographed to just "Mrs. Disney".

- A series of 30 books published by Grosset and Dunlap in from 1952 through 1959) is on the shelves, with such titles as *The Story of Theodore Roosevelt*, *The Story of Good Queen Bess*, *The Story of Abraham Lincoln*, *The Story of Amelia Earhart*, *The Story of Marco Polo*, and *The Story of Joan of Arc*. One title was *The Story of Pocahontas* by Shirley Graham from 1953.

- *A Dragon on the Hill Road* (Valley Village Press, 1958) a narrative poem written by storyman Richard "Dick" Huemer and illustrated by Disney background artist Walt Peregoy, is on the shelf, autographed by Huemer to Walt.

There weren't just books on the bookshelves. On the top of one bookcase was a metal pagoda on a wooden base, roughly 16 inches by 10.5 inches, labeled "Presented to Walt Disney for his outstanding achievement in fostering international understanding through photography under the auspices of P.P.A. by Nippon Hogaku, K.K. 1961".

On top of that same bookcase was a bronze metal bust of Abraham Lincoln mounted on a marble base labeled "Commemorating the opening of the Illinois Pavillion, New York World's Fair, 1964–65", produced by Alva Museum Replicas, Inc.

Walt Disney had a great curiosity about a wide variety of things. The huge diversity of the books in his libraries reflects that interest and are a glimpse into what inspired and intrigued the master storyteller.

PART TWO
Disney Film Stories

In January 1959, film historian Tony Thomas interviewed Walt Disney. Thomas joked, "We don't think of you in terms of the silent picture era."

That statement remains true today as well. Of course, one of the things that Walt was famous for was the introduction of synchronized sound in animated cartoons with *Steamboat Willie* (1928) that introduced the character of Mickey Mouse. He was also well known for the first three-strip Technicolor cartoon, the Silly Symphony *Flowers and Trees* (1932), that won the first-ever Oscar for an animated short cartoon.

So, understandably, most people think of Disney cartoons with sound and color. Yet, Walt's career in animation began in the age of silent black-and-white cartoons, and he was one of the pioneers of that art form.

During the famous September 25, 1963 Canadian television show interview with Fletcher Markle, Walt said:

> When Mickey [Mouse] appeared, I was making cartoons long before that. In fact, I think I've been in the business as long as anyone living today. Of course, they were very crude things then... We didn't draw them like we do today.

In 1922, Walt produced seven Laugh-O-Gram animated shorts that updated classic fairy tales like "Little Red Riding Hood" and "Cinderella" into the 1920s. When his company went bankrupt, he moved to Los Angeles where he was responsible for another silent black-and-white animated series, the Alice Comedies. The premise of those cartoons was a six-year-old live-action girl interacting with cartoon animals and a cartoon background. The series was so popular that fifty-seven cartoons were made from 1924–1927.

Then, beginning in 1927, Walt produced a series of cartoons featuring Oswald the Lucky Rabbit. His loss of that cartoon superstar is what spurred Walt to create Mickey Mouse in 1928.

The story of Walt's earliest silent black-and-white animated triumphs are often forgotten, so this section includes the complete story of the groundbreaking silent animated short *Gertie the Dinosaur* (1914) that inspired him. In addition, there is a chapter devoted to his first silent cartoon superstar, Oswald the Lucky Rabbit.

The Disney films are really the stories of characters and people whose hard work captured the hearts of many generations. This section also includes behind-the-scenes stories of some of them, including one non-Disney produced film starring actor Tony Curtis that has been the focus of much speculation over the years.

Jiminy Cricket Is No Fool

I'm no fool! No sir-ee!
I wanna live to be 23.
I play safe for you and me
'Cause I'm no fool!

Anyone can be a fool
And do things which are wrong
But fools find out when it's too late
That they don't live so long.

— *I'm No Fool* (1955)
Words and music, Jimmie Dodd

During the following choruses, the age increased from 23 to 33, then skipped to 53, 93, and finally 103. This was the memorable theme song sung by Jiminy Cricket (by his original voice artist, Cliff Edwards) for a series of animated shorts that aired on the original *Mickey Mouse Club* television series in 1955.

Walt Disney was no fool when it came to the production of *The Mickey Mouse Club*. He realized that doing the hour-long show would guarantee at least another $1.5 million from ABC to help finance the construction of Disneyland. However, it was important that costs be contained on the children's show. Besides the opening credits that could be recycled every day, additional animation could be created only if there were other ways to help recoup those production costs.

Since *The Mickey Mouse Club* was to include an educational approach as part of the entertainment, Walt realized that he could create animated educational shorts that could then be rented to schools and other civic institutions, like police and fire departments, through the Disney Studio's new 16mm Film Rental Division. (That division evolved into the subsidiary Disney Educational Media Company in 1969.)

So, even though the cartoons originally aired in black and white on a small television screen, they were filmed in color so they could be more appealing when shown in those other venues.

In the third issue of Walt Disney's *Mickey Mouse Club Magazine* (Summer 1956), there was a two-page color adaptation illustrated by artist Paul Hartley, with the last paragraph stating:

> These illustrations are from the Walt Disney film *I'm No Fool With Fire*. This eight-minute Technicolor film may be rented if your school, Scout troop or any other organization you belong to wants to show it. Write 16mm Division, Disney Studio, Burbank, California.

Many schools took advantage of that offer, as well as renting other 16mm films (including complete Disney animated shorts and features) for fundraising purposes.

On the evening of September 23, 1955, just ten days before the debut of the original *Mickey Mouse Club* on television, Walt Disney in a coast-to-coast eighty-two station closed circuit broadcast from the ABC studios in New York City, described the show and the philosophy behind it.

At one point, he said:

> Our old friend, Jiminy Cricket, will also be part of the show. Jiminy's going to help us with what we call our "factual entertainment". He'll show the youngsters things about the living world, about health, hygiene, safety, and many other things that concern their well-being.

The *Disney on Television Classroom Guide* was distributed by Disney and ABC to teachers throughout the country with program notes, guide sheets suggesting classroom activities, and other things: "A supplementary materials guide for classroom use in connection with two television programs produced by Walt Disney: *Disneyland* and *Mickey Mouse Club* presented exclusively over the ABC Television Network."

In the 1955 edition, Walt wrote:

> We have the greatest respect for the basic intelligence of our future adults and their desire to learn. We, likewise, are aware of a sometimes prevalent habit of 'talking down' to audiences of this type. To the best of our ability we aim to 'talk up' as much as possible as we program our material, remembering that we will accomplish more if we 'entertain' as we go along...Jiminy Cricket, who was rather proud of his work as Pinocchio's conscience, plays an active part in several portions of the *Mickey Mouse Club*.

A series of four different series hosted by Jiminy Cricket would deal with biology (*You: The Human Animal*), safety (*I'm No Fool*), wildlife (*The Nature of Things*) and general knowledge (*Encyclopedia*).

Another Jimmie Dodd written song sung by Jiminy is also part of many Disney childhoods:

> Get the Encyclopedia!
> E-n-c-y-c-l-o-p-e-d-i-a. Encyclopedia!

If you want to know the answers, here is the way.
Just look inside this book and you will see,
Everything from "A" clear down through "Z".
In the Encyclopedia! E-n-c-y-c-l-o-p-e-d-i-a.

Film critic Leonard Maltin wrote:

> Didn't everyone who watched the Mickey Mouse Club learn to spell "encyclopedia" that way? That's the way I spell it to this very day, with that same melodious cadence in my head.

Encyclopedia covered, in animation and live action, many topics, including a history of milk ("The final step is bottling. Boy, look at those caps go on! Yes, sir, milk is good to drink and it's made into cheese and butter and ice cream and...oh, lots of things. And who do we have to thank for all this? Bossy, here. Y'know, with all his knowledge, man has never been able to make a machine to replace the cow."), America's railroads, steel, and cork.

The *Encyclopedia* opening was later adapted so that Jiminy introduced some of the MMC Newsreel Specials, like the visit to Washington, D.C. In addition, Jiminy hosted the Mickey Mouse Book Club, which was a way to use clips from Disney films while supposedly promoting books like *Secrets of Life, Cinderella, Uncle Remus, Lady and the Tramp,* and *The Littlest Outlaw.*

The Nature of Things taught audiences about animals. The majority of this segment generally featured live action snipped from the True-Life Adventures theatrical series with voice-over narration.

According to the Disney guide:

> In this series, Jiminy assumes a role as a moderator or master of ceremonies, as he discusses and sings about a number of things which interest him. Specifically, he is fascinated with the "case" histories of some of our most popular animal friends.
>
> With animation and photography and a special song, he delves into the background, the characteristics, and the special qualities of the Horse, the Beaver, the Elephant, the Giraffe, and the Camel. He shares with his audience the realization that many things associated with these animals which appear strange are really just *The Nature of Things.*

You (titled *This is You* in the original proposal and sometimes referred to as just *The Human Animal*) dealt with the human body, including the five senses and the proper food put into the living machine.

According to the 1955 Disney guide given for classrooms:

> Jiminy finds a discussion of the human body and how it works to be not only a rich source for his particular brand of humor, but a subject full of never-ending marvels. He limits himself in this series to spotlighting the five sense of man and the relative degrees to which each has been developed."

The same "Y-O-U" chalkboard character used in the *I'm No Fool* series also pops up here to be taught about the human body:

> You are a human animal. You are a very special breed. For you are the only animal who can think, who can reason, who can read! Now all your pets are smart. That's true. But none of them can add up two and two. Because the only thinking animal is you, you, *YOU!*

But the Jiminy Cricket segment that made the strongest impression on young viewers was *I'm No Fool.*

The series mimicked the format of Walt's first live-action film, *Tommy Tucker's Tooth* (1922), about proper dental care. The good character does everything properly and is contrasted with another character who sets a bad example and his actions are often so outrageous that they provoke laughter.

The Disney guide stated:

> Each unit is in full animation and utilizes humor and a catchy song with varying lyrics to fit the situation. As the title of the series implies, Jiminy believes it's smart not to be foolish.

The opening of the series was always the same animation, seamlessly transitioning into the new animation needed for each episode.

Sitting comfortably on the loop handle of an antique candle chamberstick and surrounded by towering mountains of enormous, different-colored books, Jiminy Cricket would burst into song and climb a stack of uneven books:

> There are two ways to do anything. The right way and the wrong way. If you want to be right, do things the right way—because if you do things the wrong way, that's the foolish way. Only fools do things the foolish way—which is the wrong way. Right?

Jiminy would leap into the air, with his open umbrella slowing his descent, until he landed by a huge blue book with the title of that day's topic prominently displayed at the top. When Jiminy opened the book, the images on the illustrated pages would spring to limited animated life.

After the brief historical background lesson on the subject, Jiminy was shown standing next to a standard school chalkboard to illustrate his lessons. Carefully, he would write the word "Y-O-U" and the chalk letters would swirl into the form of an idealized young boy who always knew the right thing to do without being told.

Later, Jiminy would write the word "F-O-O-L" and it would squiggle into a goofy-looking young boy who was always foolish and reckless and whose adventures end in disaster, turning him into chalk dust to be scooped up by Jiminy's eraser. This character was the "Common Ordinary Fool".

The animation for these chalkboard "stick figures" was done by under-rated animator Cliff Nordberg.

At the end of various examples of the right and wrong things to do in a sort of "safety game" between You and the Fool, Jiminy would pin an "I'm No Fool" button on the smart lad. "The winner and still champion!" Jiminy proclaimed.

The shorts were directed primarily by Bill Justice, who often did double duty by doing some of the animation as well. Les Clark also did some direction, too. Layout was by X. Atencio, who also did some animation and was responsible for the huge head title card. Story work was done by Bill Berg and Nick George (a former assistant of Norm Ferguson). Animation was also done by Fred Hellmich, Al Coe, John Freeman, Jack Parr, Phil Duncan, Bob Carlson, Jack Boyd, Jerry Hathcock, Bob Youngquist, Cliff Nordberg, and George Nicholas among others.

Disney Legend Floyd Norman told me:

> I did work on *The Mickey Mouse Club* as my first assignment as a Disney apprentice in-betweener back in 1956. I worked for Rolly Crump, the assistant animator who would eventually be known as an Imagineering legend. We did the Jiminy Cricket stuff for animators such as Jack Parr and Bob Carlson.
>
> We often saw Cliff Edwards upstairs in [director] Les Clark's unit. I worked with Rolly on the *I'm No Fool* series and the *Encyclopedia* series as well. As a matter of fact, I still have a few sketches of Jiminy Cricket I did back in 1956.

During the original run of *The Mickey Mouse Club*, five eight-minute *I'm No Fool* shorts were produced:

I'm No Fool...With a Bicycle
Original premiere: October 6, 1955

Jiminy Cricket gives a short history of the bicycle and then shows basic safety rules for riding. This short premiered on the fourth episode of the original Mickey Mouse Club. Released in 16mm for rental on April 1956. Updated version September 1988.

I'm No Fool...With Fire

Jiminy Cricket gives a history of man's discovery and reliance on fire through history and shares some lessons on how to properly handle its potentially destructive nature. Released in 16mm for rental on April 1956. Updated version September 1986.

I'm No Fool...As a Pedestrian
Original Premiere: October 8, 1956

Jiminy Cricket shows the history of reckless driving from 3000 AD up to the present, and then illustrates the problems faced by modern pedestrians and how to walk safely in an area with traffic. Released on 16mm for rental on October 1956. Updated version 1987.

I'm No Fool...In Water
Original Premiere: November 15, 1956

Jiminy Cricket shows the proper way to behave while swimming, and basic water safety rules. Released on 16mm for rental on April 1957. Updated version 1987.

I'm No Fool...Having Fun
Original Premiere: December 17, 1956

Jiminy Cricket discusses the history of recreation and emphasizes the rules of safe recreational fun. Released on 16mm for rental on April 1957.

Fifteen years later, a sixth *I'm No Fool* cartoon was made. Since Edwards had passed away in 1971, he was unable to record new material. Snippets of his voice, including the opening song, were used, but new material of Jiminy talking was supposedly done by Sterling Holloway. The voices did not match at all.

I'm No Fool...With Electricity
Original release date: October 1973

Jiminy Cricket discusses the discovery and history of electricity and gives rules for avoiding electrical accidents by respecting electrical safety rules. Updated version September 1988.

In the late 1980s and early 1990s, Disney Educational Media created new additions to the series, as well as updating some of the existing cartoons. Eddie Carroll, the official voice of Jiminy Cricket since 1977, supplied the voice for Jiminy when any new dialog was necessary.

Generally, these versions used the original opening song and history lesson, and then transitioned to live action, eliminating the "You and Fool" segment. New characters like Pinocchio, Gepetto, and an outer space alien were sometimes included. These versions were sometimes twice the length of the originals, with the additional live-action, and could be followed by a live question-and-answer session, often conducted by a police officer or a teacher. The suggested age level was kindergarten through second grade.

While the updated versions offer more current information, they lack the simple charm and effectiveness of the original episodes.

I'm No Fool With Fire [updated]

Original Release Date: 1986 (9 minutes)

New material includes live action of a young school boy and a young girl sharing with a fireman named Captain Brody what they know, including "stop, drop and roll", and creating a home exit route in case of fire. They are writing reports for their school on fire safety and are given a tour of the fire station. The updated version does not include the animated segment with scenes of "You and the Fool", eliminating Jiminy Cricket dialog like "Try it again, stupid." The new live-action material was produced and directed by Martha Moran with a special thanks to Ed Reed of the Los Angeles Fire Department.

I'm No Fool as a Pedestrian [updated]

Original Release Date: 1987 (15 minutes)

Jiminy Cricket introduces the program, which stars Geppetto and Pinocchio, who is learning how to be an "Expert Pedestrian". With the help of his friends Billy, Amy, and Maria, Pinocchio learns many important guidelines for pedestrian safety, from the "stop, look and listen" tool to making good decisions about crosswalk signs.

I'm No Fool in the Water [updated]

Original Release Date: 1987 (9 minutes)

Jiminy Cricket introduces this program on water safety with a light-hearted overview of primitive man's first contact with water taken from the original short. Next, Jiminy makes viewers aware that there are wrong places to swim. Connie, the lifeguard, then takes over and teaches three youngsters some important water safety rules. With Connie's coaching, the children practice the survival float maneuver and then excitedly—and safely—swim off for a race.

I'm No Fool with a Bicycle [updated]

Original Release Date: September 1988 (16 minutes)

Jiminy Cricket introduces the program, which stars Geppetto and Pinocchio, who is learning how to ride a bicycle safely. With the help of his friends Eric and Denise, Pinocchio learns safety rules for riding his bicycle and where to ride his bike. He also learns how to make his bike safe to ride and how to properly dress and wear a helmet.

I'm No Fool...With Electricity [updated]
Original Release Date: September 1988

I was unable to find information on the updated version, although it probably followed the format of the previous films of eliminating the "You and the Fool" sections and adding live action.

I'm No Fool...In Unsafe Places
Original Release Date: January 1991 (14 minutes)

One of the lessons Pinocchio has to learn after becoming a real boy is how to recognize safe and unsafe play areas. Unsafe places include railroad tracks, crosswalks, pools, and storm drains, and playing in abandoned refrigerators or at construction sites. There is also a 28-minute version available.

I'm No Fool...On Wheels
Original Release Date: January 1991 (13 minutes)

Pinocchio learns the proper procedures and equipment for riding bicycles, roller skating, and skateboarding from his live-action friends. There is an expanded 25-minute version available, as well.

I'm No Fool...With Safety at School
Original Release Date: January 1991 (12 minutes); an expanded version, March 1993; runs for 28 minutes)

Jiminy Cricket and Pinocchio show how to behave safely at school with their live-action elementary school friends.

I'm No Fool...In a Car
Original Release Date: April 1992 (15 minutes)

Proper automotive safety is explained after an alien unbuckles his seat belt in his spaceship and falls to Earth, where he learns about car safety.

I'm No Fool...In an Emergency
Original Release Date: April 1992 (13 minutes)

A police officer gets injured during a chase trying to capture an outer space alien. Two live action children demonstrate how to handle the situation calmly by calling the paramedics.

I'm No Fool...In Unsafe Places II
Original Release Date: April 1992 (15 minutes)

An outer space alien learns how to keep away from unsafe and hazardous areas with the help of two live action children.

These updated editions are no longer available from Disney Educational Media, not only because the information may be out of date, but the clothing styles and acting in the live action sections look odd to today's audiences.

In *Pinocchio* (1940), the Blue Fairy had dubbed Jiminy "lord high keeper of the knowledge of right and wrong, counselor in moments of high temptation, and guide along the straight and narrow path" so he was the natural choice to host these educational segments for children to give them the necessary advice to live happily ever after.

For a generation of young people, these shorts are how Jiminy Cricket is best remembered. The Disney Company currently uses Jiminy as the mascot of their Environmentality program, where he educates cast members and guests about the importance of taking care of the environment.

The Making of the Original "Frankenweenie"

Once upon a time there was troubled young boy who loved monster movies and lived in a bright, clean suburban neighborhood. He dearly loved his playful dog, but felt estranged from his own parents. He spent his spare time not associating with his schoolmates, but creating home-made 8mm movies.

That was both the real-life story of legendary filmmaker Tim Burton and the storyline of one of his early short films, *Frankenweenie* (1984). The original black-and-white, live-action version of this film was inspired by Universal's classic *Frankenstein* (1931), not only in terms of plot, but in visual conception, as well.

A young boy named Victor Frankenstein (played by Barret Oliver) finds release for his alienation from his suburban neighborhood by making 8mm home movies sometimes featuring his beloved pet bull terrier, Sparky.

After Sparky chases a ball into the street and is hit by a car, the heartbroken boy learns in his school science class how electrical impulses can result in muscle movement thanks to a demonstration on a dead frog. (Or as teacher Paul Bartel says, in an homage to the famous Monty Python dead parrot sketch, an "ex-frog".)

The boy converts his attic into an elaborate makeshift laboratory and, just as in the classic film, uses the electrical power in a bolt of lightning to revive the corpse of his dog. While young Victor is joyful at his success, his neighbors are terrified of the stitched together canine with bolts sticking out of his neck, an homage to Jack Pierce's iconic make-up for Boris Karloff in *Frankenstein*.

Running away from these upset neighbors, Victor and Sparky find themselves at the local miniature golf course and hide in the windmill structure. An angry mob accidentally sets the windmill on fire and an unconscious Victor is rescued by his dog, who dies again in the attempt.

The repentant mob use their cars and jumper cables to re-charge Sparky and bring him back once more to life. It is a happy ending for Sparky, who falls in love with a poodle whose hairstyle resembles the classic hairstyle of the monster's bride from the film *Bride of Frankenstein*.

At the age of 25, Tim Burton directed *Frankenweenie*. It was an approximately 30-minute black-and-white short costing the Disney Studio about one million dollars.

When the film was finished, Burton remarked:

> *Frankenweenie* came out of some drawings and some feelings, and then thinking maybe this could be good, maybe we could do this as a featurette... You have a dog that you love, and the idea of keeping it alive was the impulse for the movie..."

In 1985, Burton told interviewer Michael Mayo:

> I had just seen *Frankenstein* again and started thinking for some reason about a dog I had when I was young," Burton told writer Michael Mayo in 1985. "I started thinking just how incredible the whole idea of Frankenstein really is, of bringing something dead back to life. But all the versions of it so far have just dealt with the horrible aspects of the idea.
>
> At some point, the idea of my dog and Frankenstein just connected and we started developing it. I put the idea on storyboards and pitched it to Richard Berger [then production chief at Disney] and he liked it. We got a writer named Lenny Ripps to write the script and continued to develop it from there.

Ripps, a well-known comedy writer, had written everything from the infamous *Star Wars Holiday Special* (where he learned from George Lucas that Han Solo was married to a Wookie wife and had been raised by Wookies as a child). to eight episodes of the TV show *Bosom Buddies*, to material for comedians Redd Foxx, Joan Rivers, Rodney Dangerfield, and others. He is still a prolific comedy writer and some feel it was his writing that brought a warm softness to the strange, macabre world of Burton's concept.

Ripps told an interviewer in 2000:

> I learned that essentially everything you write is autobiographical, from something you experienced, felt, or perhaps just observed. I always begin by writing about things that happen to actual people, then funny can follow. Somehow, for me, going from being funny first is backwards.

Unfortunately, it is a sad reality that we all outlive our four-legged childhood friends and that final goodbye can be extremely painful, as Burton remembered about his own dog from his early days in Burbank:

> His name was Pepe—we lived in a Spanish neighborhood. Our dog had this thing called distemper, and wasn't supposed to live more than a couple of years. He lived much longer than that, which kind of fed into this Frankenstein mythology as well. He was a mix, kind of a mutt, with a bit of terrier, and a bit of something else. I don't know what it was. It was kind of a mixture."

It's such an unconditional relationship. A lot of kids have that experience – I certainly had that experience with a first pet. You'll probably never have it again in your life in that way, it's so pure and memorable. What's more pure than the story of a kid and his first pet? Mix that with the Frankenstein myth and it causes problems. Ultimately, we try to go with the slightly more positive aspects of keeping that [boy-and-his-dog] relationship going.

Richard Berger asked an executive in the story department, Julie Hickson, who had worked with Burton on his aborted *Trick or Treat* project, to produce the film. In an interview with Mayo in 1985, Hickson said:

I think that if you look at Tim's drawings, aside from the artistry involved, there's a lot of ideas there...they're really jam packed, and it's exciting to work for someone like that... We did a version of *Hansel and Gretel* for the Disney Channel that didn't turn out to be a big hit. It was a candy-land martial arts version of the story with an all-Oriental cast that didn't have a big budget, but we had a lot of fun doing it.

It was Hickson who approached actress Shelley Duval, whom she knew casually because Duval had tried unsuccessfully to sell Disney on her idea for a series to be called *Faerie Tale Theatre*. Hickson felt that if she got the material to Duval directly, the actress would respond to it. Actor Daniel Stern came on board because he had wanted to work with Duval. Actor Paul Bartel, always a strong supporter of small independent films, loved the material, as well. Hickson recalled: "Basically, we got this cast for no money because they all wanted to do it."

A decade after the film's release, Burton told another interviewer:

They [the actors] were all great. All of these people, they knew I had never done anything before, but they liked the idea [of the film]. They felt that I cared. I think what they did was make me feel comfortable and I started to learn that you have to communicate with people.

The final cast was Shelley Duval (Susan Frankenstein), Daniel Stern (Ben Frankenstein), Barret Oliver (Victor Frankenstein), Joseph Maher (Mr. Chambers), Roz Braverman (Mrs. Rose Epstein, who has a dachshund named Raymond), Paul Bartel (Mr. Walsh, the science teacher), Sofia Coppola (Ann Chambers, credited as "Domino"), Jason Hervey (Frank Dale), Paul C. Scott (Mike Anderson), Helen Boll (Mrs. Curtis), Bob Herron (Street Player), Donna Hall (Street Player), Rusty James (Raymond), and Sparky as Sparky.

The original Kenneth Strickfaden electrical equipment for Universal's *Frankenstein* was used in the attic laboratory. It had been in storage for decades, but was found and used in Mel Brooks' *Young Frankenstein* (1974) about ten years earlier. "It was hard for Disney to understand why we wanted the equipment so badly until they saw it," Hickson said.

Rick Heinrichs, who produced Burton's stop-motion short, *Vincent*, was the associate producer on the film. *Vincent* has a short moment foreshadowing *Frankenweenie*, where a mad doctor version of Vincent is wiring up his dog to electrodes in an imaginary attic laboratory.

There was only enough budget for about two weeks of pre-production time for *Frankenweenie*, since Disney was trying to keep production costs down because of all the studio overhead. "The actual production was a fifteen-day shoot with a couple of months for post-production," Hickson said.

Burton initially complained about shooting a black-and-white picture on color film stock, but the final film has strong, crisp contrasts between light and shadow.

In later interviews, Burton has tried to explain how he utilized the connection between the classic Universal horror film and his short film. Interestingly, the lead character in Burton's film is named Victor, just as in Mary Shelley's original novel. (The Universal film renamed the character "Henry".) Young Victor's breakfast mug is in the shape of Lon Chaney Jr.'s head as he appears in Universal's *The Wolf Man*. Right next to that mug is a carton of Donald Duck orange juice.

As Burton remembered in the book *Burton on Burton* (1995):

> We did *Frankenweenie* as if the original story had never existed. This suburban family is the Frankenstein family and the little boy is Victor, but it's not a nudge-in-the-ribs type of thing. We don't have the family watching the Universal original as a foreboding of things to come. I don't think this is a dark or macabre story, and we didn't try to make the dog something horrible. He brings the dog back to life because he really loves the dog.
>
> For some reason, I was always able to make direct links, emotionally, between the whole Gothic/*Frankenstein*/Edgar Allan Poe thing and growing up in suburbia... Growing up in suburbia there were these miniature golf courses with windmills which were just like the one in [the big climatic scene of] *Frankenweenie*. There were poodles that always reminded you of the bride of Frankenstein with the big hair. All those things were just there. That's why it felt so right or easy for me to do—those images were already there in Burbank.
>
> It's very, very important to me, even though there are feelings from *Frankenstein* that I do not make direct linkage to it. If I was to sit down with somebody and we were to look at a scene from *Frankenstein* and say "Let's do that," I wouldn't do it, even if it's an homage or an inspired kind of thing. I try to make sure in my own mind that it's not a case of "Let's copy that"... The writer, Lenny Ripps, was that way. He got it. He didn't want to sit there and go over *Frankenstein*. He knew it well enough. It's more like it's being filtered through some sort of remembrance.

For *Frankenweenie* I didn't look at anything. I remember thinking the skies in *Frankenstein* were really cool because they were painted. But I didn't go and look at the film because I didn't want to say, "Do it like that." I wanted to try to describe it the way I remembered. So I would describe something, and say, 'It was like a painted backdrop, but the clouds were more pronounced. It was a much more intense, wild sky.' Then when I finally looked at Frankenstein, I saw that the sky was not quite the way I had described it. That was my impression, but I would still rather go with that. I feel when somebody is just borrowing something, they don't have any feeling for it themselves.

I never considered myself a writer, even though I do write things. I feel whether or not you write it, you have to feel like you wrote it. I could see it a little more clearly by having somebody else write it. I've always felt as long as they get me and get what it is that I feel, then they can bring something to it themselves. Then it's better. It opens it up a little bit more.

Originally, the film was scheduled to go into production early in 1984 to play with the summer 1984 re-release of *The Jungle Book*. Disney decided instead to delay shooting until later summer and have the film be shown with the Christmas re-release of *Pinocchio* in 1984. *Pinocchio* was Hickson's favorite Disney animated feature and Burton was excited, as well.

"We'll have a beautiful black and white film with one of the best color movies ever made," Burton said. "When I look at stuff we asked them to make from the designs, I don't think we could have gotten it from any other studio."

Two test screenings in September 1984 with mothers and young children resulted in the short earning a PG rating from the MPAA (Motion Picture Association of America). Mothers were concerned that the film would encourage their children to play inappropriately with electricity, among other things, as well as the general "intensity" of the film.

Burton said:

It freaked everybody out that *Frankenweenie* got a PG rating, and you can't release a PG film with a G-rated film. I was a little shocked, because I don't see what's PG about the film: there's no bad language, there's only one bit of violence, and the violence happens off-camera. So I said to the MPAA, "What do I need to do to get a G Rating?" and they basically said, "There's nothing you can cut; it's just the tone." It was supposed to be released with *Pinocchio* and I think *Pinocchio* has some intense moments. I remember getting freaked out when I was a kid and saw it. I remember kids screaming.

The film had a short limited theatrical run in Los Angeles in December to qualify for Academy Award consideration. It did receive a small release in the U.K. on a double bill with Touchstone Pictures PG-rated *Baby: Secret of the Lost Legend*, in 1985. The film finally appeared on videotape in 1992 to take advantage of Burton's new reputation as a big box office filmmaker. The short is one of the extras on the *Nightmare Before Christmas* DVD.

Burton recalled:

> It was right at the time when the company was changing [Michael Eisner and Frank Wells were now in charge]. I remember being frustrated that the old regime was out, the new one was in, and again a 30-minute short is not a high priority. [The feeling was] "Oh, this is great but we have no plans to release it. Ever." By that point I was really tired of Disney. It was a case of doing a bunch of stuff that nobody would ever see. It was kind of weird.

Of course, a new administration had little interest in promoting the accomplishments and projects of the preceding administration.

Luckily, in 1984, Paul Reubens was looking for a director for a film idea he had been developing for many years. At a Los Angeles party, Reubens was asking around for suggestions and one of the guests had just seen *Frankenweenie*.

Horror writer Stephen King had also seen *Frankenweenie*, and strongly recommended it to Bonni Lee, an executive at Warner Brothers. Lee then showed the film to Paul Reubens. (King had covered similar ground of bringing a beloved childhood pet back to life with the release of his novel *Pet Sematary* in 1983.)

Reubens arranged a meeting with Burton and the two men immediately bonded.

Some of the folks involved with the production of *Frankenweenie* ended up in Burton's first full-length live-action film, *Pee-Wee's Big Adventure*: Bob Herron did stunt work, Sandy Berke Jordan did costuming, Christy Miele coordinated the animals, and Jason Hervey played the spoiled-brat actor Kevin Morton. The film opened up a new creative career for Burton as a director of live-action features.

Burton had always wanted to expand *Frankenweenie* to feature length. As early as November 2005, development work was being done and possible scripts written. The expanded film was again shot in black and white as well as Digital 3D, and released by Disney in October 2012. It received good reviews and grossed over eighty million dollars worldwide as well as being nominated for an Oscar as Best Animated Feature, which it did not win. The film was released on DVD and BluRay in January 2013.

In July 2012, Burton said:

> It is a project that always meant something to me. But, the opportunity to do it stop-motion, in black and white, and expand on it with other kids and other monsters and other characters, it just seemed like the right medium for the project.

I just remembered some of the kids that I went to school with and some of the weird teachers. And growing up with those Universal horror films, like House of Frankenstein or House of Dracula, where Frankenstein met the Wolfman and they combined them together, a lot of it had to do with those kinds of things that I love. Even though it's revisiting something that I did a long time ago, it feels new and special.

The Secret Story Behind "40 Pounds of Trouble"

One non-Disney movie that has been both fascinating and frustrating for many Disney fans is the Universal film *40 Pounds of Trouble* (1962). It was rarely available for viewing after its initial theatrical showing (except for a few occasional edited airings on television). It did have a brief release on VHS in 1996, but quickly disappeared.

Very little information exists on this unusual film, and there have been many questions about its Disney connection over the years. Of the movie's roughly 106 minute running time, nearly 20 of those minutes are full color footage of Disneyland, from a ride on the Matterhorn bobsleds to a visit to Tom Sawyer's Island to some rather odd Disneyland experiences that never happened for any regular guest.

The short story "Little Miss Marker" by journalist Damon Runyon (two of his other short stories were combined into the musical *Guys and Dolls* and another inspired the film *Pocketful of Miracles*) tells the tale of five year old Martha "Marky" Jane whose father leaves her in a gangster-run gambling establishment as a "marker" (collateral) for a bet. When the father dies before he can return to claim her, the gangster Sorrowful Jones is left with the kid and, with the help of a sentimental floozy, takes care of the girl.

The story was made into a Paramount Pictures movie with Shirley Temple (the role made her a box office star) and Adolphe Menjou in 1934. It was remade in 1949 by Paramount as "Sorrowful Jones" with Bob Hope and in 1980 by Universal as "Little Miss Marker" with Walter Matthau.

What makes this information of interest to Disney fans is that it was also remade by Universal as *40 Pounds of Trouble*, starring Tony Curtis. That version features rare full-color footage of Disneyland in 1962, showcasing some things that would disappear completely over the following years, as well as many surprises. It was the first and only non-Disney produced theatrical film that was ever filmed at Disneyland while Walt Disney was still alive.

Originally filmed in Eastman Color and Panavision, the current DVD release has a good 16x9 widescreen anamorphic transfer so it looks more appealing than it has in years of poor dubs from existing VHS copies.

Steve McCluskey (Tony Curtis) is the manager of the Villa D'Oro, a plush $25 million gambling establishment just over the California line at Lake Tahoe, Nevada, owned by Bernie "The Butcher" Friedman (Phil Silvers).

(The Lake Tahoe sequences were shot at Harrah's Lake Club, at the south end of the lake, that had opened in June 1955. This was one of the first films to show the actual interior of an operating casino. The Lake Club was sold to Harvey Gross in 1956 and renamed Harvey's Lake Tahoe.)

Steve has such a hatred for his ex-wife that he refuses to pay her any alimony, but cannot be served by an attorney for non-payment unless he crosses the state line into California, which he does occasionally to visit Los Angeles because he enjoys the challenge of the chase.

While Steve is successful, well-liked, and happy, things change rapidly with the arrival of a new husky-voiced lounge singer, Chris Lockwood (Suzanne Pleshette). Everyone including Steve thinks the young woman is the mistress girlfriend of Bernie, but she is actually his niece.

At the same time, a little six year old weighing forty pounds named Penny Piper (Claire Wilcox, in her first movie) pops up. Her father left her in his hotel room to make a fast trip to San Francisco to grab some cash to pay off his gambling losses. When the girl's father doesn't return, Steve decides to avoid adverse publicity for the casino and moves her into a room in his suite where he begins to bond with her like a surrogate father. Penny's big obsession is to go to Disneyland, but Steve keeps deflecting her requests.

Seeing his sincere attention to the youngster's welfare, Chris warms up to Steve, and Penny encourages that connection. Unfortunately, it is learned that Penny's father has not returned because he was killed when his car accidentally plunged over a cliff.

Unable to tell Penny the truth, especially since she is also motherless, Steve decides to take her to Disneyland first to help ease the news. He quits his managerial role to do so. Chris convinces him to take her along as well so that they will look like a typical family and because she can do things that Steve cannot do, such as take the girl into the ladies' room.

If Steve gets caught in California, he will be served papers by his ex-wife and arrested. Of course, Steve's ex does find out and the result is a wild chase through Disneyland by an unnamed cigar chomping police detective played by 6'3" tall actor Tom Reese, who always seems one jump ahead of Steve. The detective is dressed in a suit and tie and wears a hat and doesn't seem the least bit out of place in the theme park.

Steve is finally caught in the Disneyland parking lot. Both Steve and Bernie are brought by the district attorney to a court hearing. The hearing takes some strange twists and Steve's request to adopt Penny is denied because of his single status.

For the happy ending, Steve marries Chris and they adopt Penny. Bernie has his gambling license renewed and is so pleased that he offers Steve his job back after Chris, Steve, and Penny return from their honeymoon... at Disneyland.

The film is a lightweight piece of fluff common in Universal's romantic comedies of the time, like *Pillow Talk* (1959), *Lover Come Back* (1961), and *Send Me No Flowers* (1963).

The screenplay was written by Marion Hargrove, who gained fame for his light-hearted columns about becoming a soldier during World War II that were later the basis for both a novel and a film.

After the war, he wrote two more novels and numerous articles for magazines. In 1955, he moved to Los Angeles and began writing for television, including episodes of *Maverick* (1957), the pilot script for *77 Sunset Strip* (1958), *I Spy* (1966), and *The Waltons* (1975). He also wrote the screenplay for the musical film *The Music Man* (1962), for which he won the Writer's Guild of America Award. That film was released the same year as *40 Pounds of Trouble*.

Norman Jewison would become one of the most successful filmmakers, with credits that included *The Russians Are Coming, The Russians Are Coming* (1966), *In the Heat of the Night* (1967), *The Thomas Crown Affair* (1968), *Fiddler on the Roof* (1971), and *Jesus Christ Superstar* (1973).

But everyone has to start somewhere, and *40 Pounds of Trouble* was Jewison's first film. He had been directing the Judy Garland television special show. Actor Tony Curtis was visiting the studio during a rehearsal and was impressed with how Jewison was handling himself. Curtis suggested that Jewison should direct a feature film. Curtis had his own production company (Curtis Enterprises) and a relationship with Universal where he had been under contract, and so Jewison was given his chance.

Curtis had been one of the many celebrity guests that attended the July 17, 1955, press preview opening of Disneyland and was equally adept in both dramatic and comedic roles. The filming of the Disneyland scenes took place in May 1962, a month before Curtis was granted a divorce from his first wife, Janet Leigh.

While Suzanne Pleshette is best remembered as Emily Hartley, the sharp-tongued independent wife on *The Bob Newhart Show* (1972–1978), she had a rich performing career. The Theater Owners of America organization declared Pleshette as "1963's Most Exciting New Star", even though she had been doing stage and television work since the 1950s and her first film was Jerry Lewis' *The Geisha Boy* (1958). Lewis and Curtis were good friends. In 1963, after *40 Pounds of Trouble*, she had a pivotal role in Alfred Hitchcock's *The Birds* as a school teacher killed by crows.

Walt later saw her performing on a television show. After inquiring about her credits, he held a special screening of *40 Pounds of Trouble*, which led to him casting her in *The Ugly Dachshund* (1966) opposite Dean Jones. So, in an odd twist of fate, Pleshette's appearance in this Universal film helped spawn her Disney film career. She also appeared in Disney's *The Adventures of Bullwhip Griffin* (1967) as a singer in a saloon and again as Dean Jones' love interest in *Blackbeard's Ghost* (1968) and *The Shaggy D.A.* (1975).

Much later, in *Lion King II: Simba's Pride* (1998), Pleshette provided the voice (both speaking and singing) of the main antagonist, the lioness Zira, who was the leader of the Outsiders in the Outlands and Scar's most loyal follower. She was also the mother of Nuka, Vitani, and Kovu, one or more of whom may be the offspring of Scar, and she had raised her son Kovu to be Scar's successor.

Disney Legend Floyd Norman said:

> Being an old Disney guy, I had the opportunity to watch Suzanne at work on the Disney sound stages on several films. She was a very funny lady, and loved to have a good laugh. I still remember her laugh as she walked down the hallway of our Animation building back in the 1960s. She was a very classy lady.

After an absence from films for many years, Phil Silvers reappeared for this production and turned down the lead role of Pseudolus in the original Broadway production of *A Funny Thing Happened on the Way to the Forum* (Zero Mostel took the part) to do this film instead. It was a mistake that Silvers vocally regretted over the years. He was cast as Pseudolus in a revival of the musical in 1972 and won a Tony Award for his performance.

The supporting cast in *40 Pounds of Trouble* was a roster of talented character actors that included Larry Storch, Edward Andrews, Howard Morris, and Stubby Kaye, as well as leading men like Kevin McCarthy and Warren Stevens. Storch appeared in five Tony Curtis films because the two had developed a friendship when they met while serving together on the submarine tender USS *Proteus* during World War II.

"Wait'll You See Their Hilarious Adventures in...Disneyland!" proclaimed the poster for *40 Pounds of Trouble*, and indeed the adventures are hilarious, but not in the way you might first suspect. Those twenty minutes are almost unexplainable and have puzzled Disneyland fans.

For those familiar with Disneyland's geography, the scenes in the park are like being dropped into Alice's Wonderland, since there is absolutely no regard to the actual layout of Disneyland or how Disneyland operated. For instance, there is a prominent silver pay phone booth on the front of the walkway just off of the Hub leading to Tomorrowland. When Tony Curtis makes a call, audiences can clearly hear the whistle from the Mark

Twain steamboat and the clanging fire bell at the Main Street Fire Station. Neither of those sounds can be heard at that location.

In fact, there never was a phone booth in that location, and later in the film, it disappears completely. It was only there long enough to satisfy a need in the story. If that wasn't enough, check out the cast member there selling souvenir hats...on a folding card table. That never existed at Disneyland, either.

But let's start at the beginning of the trio's trip. They arrive by helicopter (a service that was offered in 1962 from the Los Angeles airport) and as they gaze out the window they see the Matterhorn and Monorail Blue as the scene transitions to them in the front cabin of the monorail arriving at the station. As they disembark, they run to the railing and look over and to their immediate delight see Town Square and Main Street, U.S.A.

They then board the surrey in Town Square (with a prominent sign on the ride showing it costs a dime) and as they trot down Main Street, they see a costumed Mickey and Minnie Mouse on the sidewalk waving to them and then a quartet of Keystone Cops (sometimes spelled Keystone Kops) all playing saxophones.

The Three Little Pigs costumed characters come skipping out of Tomorrowland's entrance. Next, the trio are on the Storybook Land boat attraction with a hostess in a rose-and-white dress that was the common costume. There are glimpses of Gepetto's village, Toad Hall, and Cinderella's village and chateau as well, and of course, high on the hill, the famous "dream castle", as the hostess describes it.

The group rides the Tea Cups, the Astro Jets, and the mine train through Nature's Wonderland (with the little girl now wearing mouse ears that she must have lost on the ride since they disappear for the rest of the film). This is followed by a trip on the old sky buckets through the Matterhorn.

A trip on Peter Pan's Flight is an interesting experience, since other than the flying pirate ship there is nothing seen from that attraction. However, there are horrifying scenes from Mr. Toad's Wild Ride (including the faux train crash at the end) and from the Snow White attraction (including the wicked old hag witch and her poisoned apple).

This is followed by a ride on the Matterhorn, sped up a bit to make it look faster and rougher, and before the mountain was enclosed. All three are crammed into the front section, with the seats behind them empty. When they get to the disembarkation area, they have to escape the pursuing detective so they all just jump back in to "do it again", which must have delighted the ride operators.

They then share a meal of hot dogs and soft drinks at the outdoor eating patio at the back of the Fantasyland pirate ship. To hide from the detective, they wear Halloween masks—of President Kennedy, Soviet leader Nikita

Khrushchev and Cuban leader Fidel Castro—that they somehow purchased just minutes ago at Disneyland.

The Cuban missile crisis was a thirteen-day showdown in October 1962 (this film was released in December) between these three world leaders about Soviet ballistic missiles in Cuba. Fortunately, the situation was resolved, and so the masks were meant as a humorous commentary, especially with a six-year-old girl wearing a Castro mask between bites of her hot dog.

Steve wants to leave the park, but Penny convinces him to make one last stop, at Tom Sawyer Island. And who should be hiding at the very top of the treehouse on the island? The cigar-smoking detective, who chases the trio through the island over the barrel bridge (running into a large, old, unco-ordinated grumpy lady, of course) and the suspension bridge (running into a scout troop trying to get across) and finally arriving at Fort Wilderness.

At Fort Wilderness is something that Disney fans never saw. In the middle of the day there is a flag-raising ceremony (which temporarily halts the chase) followed by Indians climbing over the walls in an attack on the fort and the sounds of rifle shots during the fight, with guests joining in to repel the savages.

The trio gets on a raft that pulls away from the island, stranding the detective while the guests on the raft all laugh as if they are in on the joke. The detective jumps into a nearby outboard motor boat yelling, "Follow that raft!"

At the entrance to Frontierland, Steve has somehow been able to obtain a full character costume from the Frontier Trading Post at the entrance—a yellow mushroom from the film *Fantasia* that was used at the time during the Christmas parade. "Stop that mushroom!" yells the detective while Chris and Penny tell Steve they will meet back up with him at the Main Street train station.

As soon as Steve exits Frontierland, he removes the costume, hops over a white fence, goes through a tunnel, and finds himself in the Living Desert (with cactus and rolling rocks and an electro-mechanical animal) and the Devil's Paint Pots (multicolored geysers; he bends over to get a drink from one that shoots off) in the Mine Train ride. He is followed by the detective, who seems flummoxed by the exploding geysers.

Steve escapes by going to the Indian Village on the Rivers of America. When Steve asks the mechanical Indian chief for directions, the character raises its arm. Steve grabs a canoe and paddles away. The detective jumps into another canoe, but his legs break through it, sinking the craft.

The canoe has taken Steve to yet another area with a tunnel and he runs in and quickly runs back out, chased by one of the Santa Fe & Disneyland railroad engines. Following those tracks leads him to Fantasyland, where he pushes the Three Little Pigs out of his way, rushes by the carrousel, and out of the castle, where he is stopped by a half-dozen Japanese sailors who want him to take their picture, which he does since there are absolutely no Disneyland cast members around.

However, the detective is hot in pursuit, and they both end up in the flat grassy Hub where they disrupt the saxophone-playing Keystone Kops, sparking a wild chase with more Keystone Kops joining in as they all race round and round the Hub in an homage to the famous silent movie comedies.

Steve runs away and hops over yet another fence to jump into a gold colored Autopia car. The detective also hops over the fence and into a car driven by a young boy and orders him to give chase. There is a brief cutaway to the Keystone Kops in a turn-of-the-century vehicle careening down Main Street with the Kops falling off, another reference to an iconic image of the silent movie comedians.

Steve jumps out of his Autopia car, runs up the hill, and jumps on the train to take him back to Main Street, where the detective is already patiently waiting. Steve grabs the little girl and runs down Main Street, encountering the White Rabbit, Goofy, and the Big Bad Wolf before buying a dozen colorful helium balloons (not the famous Mickey Mouse-headed balloons), and then running out into the parking lot where finally he is caught.

For those who miss the old Disneyland parking lot that became Disney's California Adventure, there is a long scene to enjoy.

There was absolutely no evidence of any Disneyland security as the characters leapt over barriers, entered exits and tunnels, and in general put themselves and others into danger. Not to mention manhandling the famous Disneyland costumed characters.

Walt wasn't happy. While in general the park had been presented as a fun place with some interesting things to do, it had also misrepresented what was there, and where.

Even today, there are people who believe there is a monorail station on Main Street, although they probably don't think that they will see on Peter Pan's Flight a policeman with an outstretched arm from the Mr. Toad Wild Ride attraction in Neverland instead of Captain Hook and Mr. Smee.

The press book for the film gleefully states:

> And they run into all the legendary characters who roam the grounds like The Mad Hatter, Bugs Bunny, Mickey Mouse, and Alice in Wonderland. Incidentally, all of these lovable characters have definite story point roles, which marks their first appearance in a non-Disney film.

Bugs Bunny? Bugs, of course, was a popular competitor of Mickey Mouse's in cartoons produced by Warner Bros. Alice and the Mad Hatter never appeared unless they ended up on the cutting room floor.

The Three Little Pigs, the Big Bad Wolf, Goofy, the White Rabbit, and Mickey and Minnie do make brief appearances, but for the most part they are just obstacles in the way of Tony Curtis and his attempts to escape the detective. It is a great opportunity to see these character costumes from this time period, but the characters are not used as significant "story points".

This type of sloppiness irritated Walt, but there was little if anything he could do as he was not consulted on the publicity or the editing of the film. Disneyland had only been open for a little over six years and many people had not yet visited the park, so this film would probably have confused them about what to expect when they did arrive.

Sharp-eyed viewers (or those who can pause the DVD clearly) will see glimpses in the background of the construction for the Haunted Mansion on the banks of the Rivers of America and views of the surrounding city of Anaheim, not as developed as it is today, as well as other hidden treasures behind the main action, including a few bored-looking cast members.

The press book is filled with some interesting information about the shooting of the film at Disneyland in stories designed to be reprinted in local newspapers promoting the film. Here is an excerpt of some of that information:

> Since its opening, July 18, 1955, more than 30,000,000 visitors have toured the famed amusement park and practically all of them have had the inevitable camera strapped around their necks with pockets bulging with rolls of film.
>
> So a camera at Disneyland is as common as five fingers on a hand. However, this was not the case with the battery of color cameras set up by Universal Pictures for the filming of location sequences.
>
> Walt Disney, who had turned down more than twenty previous requests from motion picture companies to use Disneyland as a background for a movie, gave producer Stan Marguiles and director Norman Jewison permission. This was the first time he has allowed his famous tourist attraction to be used for a major role in a feature length film.
>
> More than one hundred studio technicians comprising the crew utilized every nook and cranny of Disneyland for cinematographer Joe MacDonald. In fact, the more than forty-million-dollar establishment literally served as the most expensive movie set in Hollywood history.
>
> Basically, what was filmed there was an exciting chase scene in which detective Tom Reese pursued Tony Curtis from one end of the park to the other in order to serve him with a subpoena. In an attempt to elude his implacable pursuers, Tony and his companions take in all the rides and sights in the park.
>
> Even the sounds of Disneyland were captured by sensitive mikes for the twenty minute sequence that is the highlight of the film. Captured were the leisurely clop-clop of horse-drawn surreys "with

the fringe on top", the "um-pa-pa" of a band concert in Town Square, the chug-chug of "horseless carriages", happy laughter of crowds, and the steaming hiss of "Old Unfaithful Geyser".

Among the hundreds of extras used each day by Universal were one hundred children, who were required by California law to attend classes on the days they worked in the move.

An open-air schoolroom was set up in the middle of Disneyland where a half dozen teachers made with the three "R's" during the filming lulls.

As far as the kids were concerned, it was an unbelievable dream come true. They kept pinching themselves to see if it were all really so. Imagine getting paid, getting on all the rides for free, and attending school right on the grounds as a bonus!

The Hollywood caterer who ran the $37,000 traveling commissary that served hundreds of movie personnel daily figured the weekly lunch tab totaled $15,000 for meals served on the location by the company.

Five movie horses were brought to the location for some scenes. If the horse worked, it received $50, but if it just stood around waiting, the price shrunk in half. However, the group of four-legged actors earned an average of $250 weekly.

The photographing of the principals in the rides presented more problems. The problems were solved with some ingenious placing of the cameras in strategic spots. Sometimes they were mounted in cars before or behind the ones carrying the actors.

In the case of the Matterhorn bobsled ride, a small, specially built camera, complete with lights, was firmly clamped in front of the sled and set automatically to roll and record the expressions of the actors as they roared around the sharp turns.

In addition to all the camera close-ups, a helicopter shot of the entire colorful park was taken to give a theater audience a panoramic view of Disneyland.

Finally, the filming of *40 Pounds of Trouble* turned out to be one of the big attractions for the many thousands of tourists. Huge trailer trucks, vans, camera cranes, power trucks, and wardrobe trailers, of all sizes, caught the eye of the curious tourists.

Stage struck young girls forgot about the sights of Disneyland for the time being and strutted by the camera hoping to be 'discovered' by Hollywood. Autograph hunters thrust books in front of Tony Curtis or anyone else who looked like he might be somebody.

Even the utility man feels that Disneyland is a "magic place". He estimated he autographed at least 50 books before the location filming was done.

Sounds like a big mess, doesn't it? It was obviously very disruptive to the Disneyland experience. So how did Universal get permission to film in Disneyland when Walt had denied all previous requests?

The answer is Jules C. Stein, a promising eye surgeon who left that career to co-found the Music Corporation of America (MCA). In 1962, he purchased Universal Pictures and needed to make an immediate impact by revitalizing the studio.

Disney historian Sam Gennawey wrote:

> Early in the 1930s, Stein and Walt Disney became close friends. Whenever Disney was stuck trying to convince his brother Roy to fund a project, he would often threaten to take it to Dr. Stein and get the project financed.
>
> Dr. Stein was one of the earliest investors in Disneyland. When Disney considered developing a city in Florida, he consulted with Stein due to Stein's experience running Universal City.

Stein had also advised Walt and Roy about the television business in the 1950s, helping them with his entertainment connections and business savvy.

After Ellis Island in New York was decommissioned as an immigration station in the mid 1950s, Jules Stein once suggested that Walt buy the historic 27.5 acre island and build another Disneyland there.

Walt had accompanied Stein on a walking tour of the Universal Pictures studio before Stein decided to purchase the company. Stein even asked Walt whether he thought having a studio backlot tour there would be a good idea. Walt thought so. The Universal Studios tram tour (with trams designed by Imagineer Harper Goff) opened quietly in June 1964 and was a huge success.

In addition, Walt was an ardent supporter of his longtime friend's ophthalmic and philanthropic efforts, including the founding of the organization Research to Prevent Blindness. When the prestigious Jules Stein Eye Institute opened in Los Angeles at UCLA in 1966, Walt felt that a mere monetary contribution would not be enough. Walt also commissioned artist Mary Blair to create a one-of-a-kind, 220-square-foot ceramic mural reminiscent of her work on It's a Small World for the walls of the pediatric waiting room of the new institute. One of the fired clay tiles has the following inscription: "To Doris and Jules Stein With Love Walt Disney".

For those who still believe that Walt Disney was anti-Semitic, Stein and his wife were both prominently Jewish.

When Dr. Stein wanted to use Disneyland as a setting for one of the first of his newest movies, Walt hesitated only slightly in saying "yes" to his friend. Walt also assumed that this might be a new avenue to generate capital for further Disneyland expansions.

After all, how bad could it all be? Walt had sometimes filmed segments at the park for his own weekly television show, including shots for the episode "Disneyland '61/Olympic Elk" that had aired in May 1961, a year before the filming of *40 Pounds of Trouble* began.

Unfortunately, Walt saw almost immediately that the filming disrupted the guest experience. Universal had no proprietary interest in Disneyland other than as another film location. Attractions were shut down and unavailable to paying guests while filming was going on. Sections of the park were blocked off, especially with all the trailers and production crew clogging the traffic flow for cast and guests.

There was the logistics of extras and equipment, and all of this activity extended for roughly a week. The confusion undercut the storytelling at the park. More important, some guests were more fascinated by the filming of a movie than by the park itself. The negatives were far greater than any positives, such as free publicity for Disneyland and some rental fees that could be used toward improvements to the park.

It wouldn't be until 1996 and the film *That Thing You Do!* that a non-Disney film would use the park as a setting for a brief clip. *That Thing You Do!* was written and directed by Tom Hanks, who had voiced the character of Woody in *Toy Story* the previous year. That is how Hanks got permission: Disney wanted to remain on good terms with him for future projects.

In 2013, Hanks would star as Walt Disney in the film *Saving Mr. Banks* where there was a scene at Disneyland set in spring 1961. For that scene, sections of the park were redecorated to resemble what it looked like in *40 Pounds of Trouble*.

So, after many years of research, the mystery of why *40 Pounds of Trouble* was allowed to film at Disneyland is revealed.

How many Disneyland fans know of another 1962 non-Disney produced film that had a short clip of the exterior of Disneyland? In the Columbia Pictures release *The Three Stooges in Orbit* (1962), the Martians Ogg and Zogg use a raygun to destroy selected earth targets, including Disneyland.

A few years earlier, Disney had contacted Moe Howard of the Stooges about his invention of a process that could make live action look like animation. Disney had intended to use it for segments of *101 Dalmatians* (1961) for the final car chase scene. It proved to be not quite what Disney wanted.

40 Pounds of Trouble is not a significant film. If not for the footage from Disneyland, it would be long forgotten. It is merely a harmless and pleasant diversion, but a nice showcase for some actors as well as for Disneyland itself.

The film racked up its share of negative reviews, with perhaps the most famous from the influential critic Bosley Crowther of *The New York Times*, who wrote on January 4, 1963:

Every time they remake *Little Miss Marker*, the famous Damon Runyon tale about a gambler who inherits a moppet and has to take care of her, they make it a little less charming, a little more commercial and crude. That is evident in *40 Pounds of Trouble*, which opened yesterday at the Palace and other metropolitan theaters.

Although no credit is given to Mr. Runyon's sentimental conceit, in which, you will remember, Shirley Temple made her screen debut, it is clearly the inspiration for the slapdash and witless script that Marion Hargrove has knocked together for this bluntly promotional film.

It has Tony Curtis acting as a jazzy manager of a Nevada gambling club who suddenly finds himself the custodian of a left-over female child at the same time that he is wooing the snappy girl singer at the club. This, of course, is the situation in which Little Miss Marker's guardian was.

But, unhappily, Mr. Hargrove's treasury of whimsy and wit is very thin, providing nothing more than labored humor for the goons who have to help amuse the child. And his weak and probably fore-ordained solution for the dilemma is a visit with the child to Disneyland, which is photographed and plugged from every angle—including the angle of salesmanship.

Considering that the first part of the picture is pretty much an illustrated plug for Harrah's Club at Lake Tahoe, Nevada, where much of it was brightly photographed, one might reckon it a television picture, with obvious commercials built in. And considering that its first-time-out director, Norman Jewison, is straight from television, it's no wonder that it has a video look.

Mr. Curtis, Suzanne Pleshette as the singer, and Claire Wilcox as the baby-talking child are as banal as spot-commercial hawkers of headache tablets or crunchy breakfast foods.

The trouble with *40 Pounds of Trouble* is that it is just too hackneyed and dull.

Still, for fans of early Disneyland, this film is much more of a treat than a trick, and with a healthy suspension of disbelief, an entertaining experience.

The Gertie the Dinosaur Story

Why is a three-dimensional building resembling a silent black-and-white cartoon star sitting at the side of Echo Lake at Disney's Hollywood Studios?

That is a question many guests have asked themselves over the last twenty-five years or so.

Gertie celebrated her 100th birthday in 2014 and no one, including Disney, really seemed to take much notice, but without her and her loveable animated antics, animation might have remained a simple "trick film" novelty and eventually faded away like other film novelties.

A pantheon of legendary names from early animation, including Paul Terry, Max and Dave Fleischer, Pat Sullivan, Otto Messmer, Richard Huemer, Shamus Culhane, I. Klein, and Walter Lantz, among many others, have publicly admitted over the years that seeing cartoonist Winsor McCay and his vaudeville act with the animated film of his playful Diplodocus was what inspired them to get into the animation business.

In November 1914, McCay, in conjunction with William Fox's Box Office Attractions Company (the forerunner of Twentieth Century Fox), later released a short black-and-white silent film to movie theaters of his vaudeville act. It included a live-action prologue and epilogue with inter-titles that showed McCay's on-stage dialog. The film was shown sporadically over the years and, sometime around 1919, the teenaged Walt Disney and Ub Iwerks saw it in a Kansas City, Missouri, movie house and were awestruck.

In *The Hand Behind the Mouse*, co-author (and granddaughter of Ub Iwerks) Leslie Iwerks wrote:

> Ub saw a cartoon that would change his life forever. The film was *Gertie the Dinosaur* by Winsor McCay.
>
> [It] remains a revolutionary and remarkable film, even today, for its ability to generate emotional pathos from the simple animated line. The concept of combining live action and animation would become one of Ub's greatest fascinations.

McCay's film even inspired comedian Buster Keaton. For his 1923 film *The Three Ages*, Keaton told his writer Clyde Bruckman:

Remember *Gertie the Dinosaur*? ... The first cartoon comedy ever made. I saw it in the nickelodeon when I was 14 [animation historian Donald Crafton pointed out that Keaton was probably closer to 19]. I'll ride in on an animated cartoon.

In the film, using stop-motion clay models, Keaton made his entrance on the head of a brontosaurus reminiscent of Gertie.

Winsor McCay was a prolific and talented artist. Today, he may be best known for the color Sunday comic strip *Little Nemo in Slumberland* (about a young boy who falls asleep and finds himself in a fanciful world) that he wrote and drew, but he also produced hundreds of editorial cartoons, advertisements, and other comic strips, as well as being a top-ranked act on the vaudeville circuit.

In 1906, he began his career on the variety stage with a routine where on a large easel and with orchestral accompaniment, he did a highly popular "lightning sketch" act where he quickly drew freehand everything from funny caricatures to his own comic-strip characters.

In addition, he developed a fascination with animation after seeing his young son's flipbooks.

McCay did not create the first animated cartoon (although he often claimed that he had), but certainly the first one that demonstrated remarkably fluid movement and characterization. McCay's lines not only moved, but established an emotional connection between the audience and the cartoon characters who seemed to think and feel like real people, especially because of their direct eye contact with the viewer.

McCay also established a process for doing animation (including key drawings, effective registration of images to prevent "jitters", and the concept of "cycling" action that reused drawings) that would later be refined by others, like Disney, into what became the standard way of producing animation.

McCay never applied for a copyright or patent on any of his techniques. He wanted others to explore using animation to tell stories. He said, "I desire no patent or copyright on it as I believe that such a process should be open to universal use, just as the discoveries in medicine are made known."

He later stated to a newspaper colleague who was urging McCay to copyright everything he had invented for the process of animation, "Any idiot who wants to make a couple of thousand drawings for a hundred feet of film is welcome to join the club."

Unlike other animators who, like magicians, wanted to keep secret the process of doing animation, McCay willingly revealed how it was done because he saw himself as a craftsman. In fact, McCay financed his animated films (including paying his later assistant John A. Fitzsimmons) from his own earnings as a newspaper cartoonist.

McCay spent much of the fall and winter of 1910 working on the 4,000 drawings needed to animate characters from his *Little Nemo in Slumberland* comic strip for his vaudeville act. He would draw a sketch and then on a huge movie screen beside him that sketch seemed to magically spring to life. He introduced this routine in April 1911 and it proved so popular he immediately began doing a second film titled *The Story of a Mosquito* (sometimes called *How a Mosquito Operates*), that was ready by January 1912.

McCay partnered with the Vitagraph Corporation of America, one of the top film distributors of the time, to do a filmed version that included a live-action framing sequence, for national release. Audiences had never seen anything like it, and some suspected it was all done with some sort of puppetry or hidden wires or even with little people in costume. McCay decided he needed to do something so outrageous that people had to accept that it was a series of drawings.

During the summer of 1912, McCay was telling interviewers that his next animated film would include dinosaurs. He told a film trade magazine:

> I have already had a conference with the American Historical Society looking to a presentation of pictures showing the great monsters that used to inhabit the earth.
>
> There are skeletons of them on exhibition and I expect to draw pictures of these animals as they appeared in real life thousands of years ago and show them as they trampled their way through dense jungles, ate a stump or pulled down a tree, or had a battle with others of their kind.

McCay was supposedly inspired by a dinosaur skeleton put on display in 1905 at the American Museum of History in New York. It was the first such creature to be reconstructed and displayed to the public and was identified as a Brontosaurus. In actuality, it should have been more properly identified as an Apatosaurus excelsus.

For reasons known only to McCay, even though he used this skeleton as the foundation for Gertie, he referred to his cartoon dinosaur as a Diplodocus, which is another sauropod genus entirely.

McCay began production in earnest on *Gertie* during the summer of 1913. He hired a next-door neighbor, John A. Fitzsimmons, a 20-year old art student, as his assistant. Fitzsimmons primary responsibility was the mind-numbing task of carefully retracing the background from a master drawing 6" by 8" onto thousands and thousands of pieces of rice paper in Higgins black ink. The rice paper was thin enough so that Fitzsimmons could easily see through it to trace. Neither of McCay's previous two films had backgrounds.

During the production of *Gertie*, McCay not only continued to draw his elaborate Sunday comic strip, but also drew for William Randolph

Hearst's newspapers' daily editorial cartoons and more than two-dozen short-lived daily comic strips like *It Was Only a Dream, As Our Ancestors Played It, According to Webster, It's Great to Be a Husband,* and *Dream of the Rarebit Fiend.*

Where did the name Gertie come from? Animator and director Paul Satterfield was an art student in Atlanta, Georgia, when he met and talked with McCay, who was performing with *Gertie* in a theater there in 1915. Satterfield relayed the following anecdote to animation historian Milt Gray in 1977:

> He told us how he happened to get the name Gertie. He heard a couple of "sweet boys" out in the hall talking to each other and one of them said, "Oh, Bertie, wait a minute!" in a very sweet voice. He thought it was a good name, but wanted it to be a girl's name instead of a boy's, so he called it Gertie.

Over the months, McCay drew roughly 10,000 drawings of the characters (by his own account) to make approximately five minutes of animation. There were no schools or books that taught animation so he had to invent a method to do animation.

McCay recalled:

> When [Gertie] was lying on her side, I wanted her to breathe and I could come to no exact time until one day I happened to be working where a large clock with a big second dial accurately marked the intervals of time.

> I stood in front of this clock and inhaled and exhaled and found that, imitating the great dinosaur, I inhaled in four seconds and exhaled in two.

McCay drew 64 drawings of Gertie inhaling and 32 drawings of her exhaling to get one cycle. To save on "pencil mileage", McCay created the animation concept of "cycling", which meant refilming the same set of drawings for a continued action. "I only drew her breathing once," McCay said, "but I photographed that set of drawings over 15 times."

Gertie made her first appearance at the Palace Theater in Chicago in February 1914 as part of McCay's vaudeville act that still included the lightning sketches and a screening of the *Mosquito* film. After the film, McCay would stand on the apron of the stage dressed in his black cutaway tuxedo, but holding a huge bullwhip like an animal trainer might use to handle a lion. He engaged in some "patter" with the audience, then drew a sketch of Gertie on a large easel and told the crowd he was going to introduce them to "the only dinosaur in captivity".

With a crack of his whip, the movie screen sprang to life and McCay coaxed the shy Gertie to poke her head from out of her cave from behind some rocks in the distance. She had the personality of a curious puppy or small child, despite her massive height and girth. As she walks toward the audience, she casually picks up a rock off the ground with her mouth and eats it, as well as gobbling down an entire tree.

McCay commands her to do a series of tricks such as lifting her legs one at a time and bowing to the audience. She is momentarily distracted by a sea serpent in the lake, and McCay cracks the whip again to regain her attention. However, she resents being pushed into performing and lunges forward to snap in his direction. Reprimanded by McCay, she bursts into tears.

To show no hard feelings, McCay offers her an apple (identified as a pumpkin in the Fox film because it would make more sense in terms of proportion, but an apple was easier to "palm" on stage to hide from the audience) that he seems to toss up at the screen.

On the screen, Gertie catches it in her mouth and delightfully devours it, then decides to lie down and take a nap during which she occasionally scratches herself with the tip of her tail.

Her rest is interrupted by a four-winged lizard flying overhead and then by a woolly mammoth named Jumbo, whom she mischievously grabs by the tail and tosses into a nearby lake. In triumph, she does a little dance that ends when the annoyed Jumbo, still in the lake, uses his trunk to spray her with water.

Gertie retaliates by picking up a large stone and throwing it at the mammoth, who is quickly swimming away. Thirsty from all this activity, she takes a drink from the lake and soon drains it completely dry with the ground beneath her at the water's edge starting to give way.

Finally, it appears as if McCay walks into the animated scene himself as a miniature caricature in tuxedo and carrying the same type of whip. He steps into Gertie's open mouth and is gently lifted and set down on her back. He bows and Gertie carries him off screen.

Supposedly, there was also a short curtain call where Gertie returned on screen and bowed again to the audience.

During the presentation, she responded to McCay and to the audience as if she were a live performer, thanks to McCay's precise timing. The film is filled with little bits of natural action, like when she licks her lips and smiles after gobbling the tree trunk. It was an interactive experience and charmed every audience.

Some hecklers in the audience were initially skeptical, as McCay reported in 1919:

> When the great dinosaur first came into the picture, the audience said it was a papier-mache animal with men inside of it and with a scenic background.

As the production progressed they noticed that the leaves on the trees were blowing in the breeze, and that there were rippling waves on the surface of the water, and when the elephant was thrown into the lake the water was seen to splash. This convinced them that they were seeing something new—that the presentation was actually from a set of drawings.

Reviewers praised *Gertie* for its originality, humor, and cleverness.

Gertie the Dinosaur might not have been officially the first animated cartoon, but it was the first animated cartoon of any consequence. Its huge success should have been the springboard for happiness and good fortune for Winsor McCay and the beginning of an amazing career in animation.

It wasn't.

One poster proclaimed it as the "Greatest Animal Act in the World", and went on to say:

Gertie: she's a scream. She eats, drinks and breathes! She laughs and cries. Dances the tango, answers questions and obeys every command! Yet, she lived millions of years before man inhabited this earth and has never been seen since!!

Gertie had a depth of personality. She was shy, stubborn, petulant, sensitive, playful, and curious, all within five minutes. Later animated characters of this time period were interchangeable ciphers. It took Walt Disney, who had been amazed by Gertie, to reinvent the concept of personality (or character) animation to this same extent many years later.

Everyone loved the act, except for one prominent person, William Randolph Hearst, McCay's employer, who felt that all of this theatrical nonsense was distracting McCay from his responsibilities for the Hearst papers, in particular the daily editorial cartoons. McCay never missed a deadline nor slacked in producing highly detailed and beautifully composed comics, but Hearst was known for wanting total control.

While Hearst could not legally prevent McCay from performing, he announced that any theater that booked the act would receive no advertising or mention in any Hearst newspaper, the most widely read newspapers of the time.

Prevented from touring with the act by theaters now afraid to book him, McCay worked with a film distributor to add a live-action prologue and epilogue for the film (more than doubling its total length by these additions). This film version of *Gertie* was copyrighted September 15, 1914.

The prologue features a wager between McCay and his fellow cartoonists, including George McManus, who later became world famous for doing the popular *Bringing Up Father* comic strip, after the car they are in breaks down from a punctured tire in front of the American Museum of Natural

History. While the tire is being changed, they go inside and see the famous dinosaur skeleton.

The film company had to receive special permission to film inside the museum and the dinosaur skeleton. They were only allowed to use natural lighting and promise that the museum would not be held up to ridicule.

McCay firmly bets a dinner for the entire group that he can bring a dinosaur to life with his drawings. Then the prologue goes on to show briefly, in an exaggerated fashion, how McCay achieved this feat, much as the prologues for his previous films had done. McCay's young son, Robert, plays his dimwitted assistant who scatters dozens of drawings on the floor.

During the showing of the cartoon at the dinner banquet, it is interrupted by title cards containing a truncated version of McCay's vaudeville stage dialog. Having accomplished bringing a dinosaur to life, McCay has won the bet. McManus pays the bill for the dinner.

However, a bogus cartoon, also called *Gertie the Dinosaur*, was released roughly a year later. It has always been assumed that it was produced by John Randolph Bray's company because of the use of animation cels to which Bray held the patent at the time.

Bray had misrepresented himself to McCay as a writer during the production of Gertie, but stole all of McCay's animation techniques and patented them himself. Then he sued McCay for using them.

In court, McCay easily proved he had been using those techniques for several years before Bray, and there was a settlement where McCay received royalties from Bray for those techniques for nearly two decades.

It is believed that Bray produced this film (without credits) to take advantage of the popularity of *Gertie* to an unsuspecting public and perhaps as a bit of revenge on McCay.

The differences between the original and the fake are obvious and significant. The background is tropical rather than prehistoric with a palm tree on the right hand side. Gertie comes out of the water as if she had been swimming underneath like a fish and onto the beach.

She is two-toned and colored gray and white rather than just white. The background in Bray's film is static and does not contain the unintentional movement of the original that seems to resemble sun shimmering on the water and a breeze going through the leaves of the tree.

Gertie moves toward the audience and rocks back and forth in imitation of the original and is interrupted by a flying lizard behind her. She shakes her head back and forth continually in disbelief and then snaps at some foliage in the foreground while wagging the tip of her tail.

She balances a coconut on the tip of her snout and bounces it up and down until finally eating it. She grabs a large rock and twists her neck into a knot and then swallows the rock, untwisting her neck.

A small black monkey climbs up the palm tree and Gertie snaps off the top of the tree to eat. The frightened monkey leaps on Gertie's neck and slides down to the ground while Gertie finishes eating. The hapless monkey gets its tail caught under one of Gertie's feet.

Gertie grabs the monkey with her mouth and tosses him into the water. The irritated monkey gets back to the beach and yells at Gertie, who picks up the rest of the tree and throws it at the monkey, who scampers away.

In the background a woolly mammoth wanders onto the beach to get a drink of water, but a sea serpent grabs him by the trunk and a tug-of-war ensues while Gertie watches. The mammoth is able to snap his trunk away. As he walks away, Gertie picks up a rock and throws it at him to speed him along. Gertie stands up on her hind legs and laughs with two of her legs holding the sides of her stomach.

A little man in a black coat enters from the right-hand side and bows to the audience. Gertie grabs him by the back of the coat and deposits him off-screen while she bows to the audience.

Unfortunately, for many decades this misidentified film was used in documentaries to represent McCay's much superior work. In the 1970s, Blackhawk released this as McCay's original "made in 1909" (sic), an error in the date that pops up frequently on the internet.

While the film has lots of action, it has none of the charm of the original or any of the artistic skill that McCay brought to showing concepts like weight and emotion.

Around 1921, McCay began work on a second animated film that was to feature Gertie. Titled *Gertie on Tour*, it was never completed, except for less than two minutes of footage showing Gertie playing with a small toad and then with a trolley car, like a cat with a mouse, until she derails it.

Exhausted, she lies down and dreams of dancing for other dinosaurs who look exactly like her. In addition, according to notes and concept sketches, she would have bounced on the Brooklyn Bridge in New York and tried to eat the Washington Monument in Washington, D.C., among other misadventures including perhaps getting stuck in a tunnel.

McCay died on July 26, 1934, of a massive cerebral hemorrhage.

In the July 1934 issue of *Scribner's* magazine, writer Claude Bragdon stated:

> It seems a pity that McCay, with his delightful fancy [in animation], should not have continued in this field which he had made his own. Walt Disney has so far eclipsed him that McCay's animated cartoons are remembered only by old-timers like myself.

By that time, two decades after the release of *Gertie the Dinosaur* in 1914, McCay had been long-forgotten by movie audiences as an animator, since he was no longer producing new animated shorts and there was no way to view his previous accomplishments. Other animators that had been inspired by his work, including Walt Disney (whose character of Mickey Mouse was at the peak of his popularity), were the ones in the spotlight, and they did remember and respect McCay.

Twenty years later, with the help of McCay's son Robert and Disney Legend Dick Huemer, who had seen McCay perform the act several times on stage, McCay's original stage routine was recreated for a segment of the Disneyland television series titled "The Story of the Animated Drawing" (November 30, 1955). Huemer told animation historian Joe Adamson:

> I saw McCay's *Gertie the Dinosaur* at the Crotona Theater in the Bronx and I was therefore able to re-create it later on a Disney television show, wherein we reenacted McCay's performance. We didn't have the expression back then, but if we had, I would have said "I flipped" when I first saw it.

During the work on the segment, Walt Disney came up to Robert McCay, gestured to the Disney Studio and said, "Bob, all this should have been your father's."

Many of McCay's animated shorts have been lost forever because they were filmed on highly flammable nitrate film that quickly deteriorated. *Gertie* was only rescued in 1947 because of a bit of luck and that there were multiple copies of it stored in a garage due to its popularity. Of the 10,000 drawings used to make the film, only 400 or so still exist in any form. Thanks to the scholarship of animation historians, like John Canemaker and Donald Crafton, McCay's work, especially *Gertie*, has been brought to the attention of the general public.

With the opening of Disney-MGM Studios in 1989, which featured an actual working animation studio, the Imagineers wanted to include a significant reference to classic animation history, in addition to the usual Disney images.

Not only was *Gertie* considered the true birth of character animation, the character and film were both in public domain. So, a full-sized version was constructed (like a building, because just below the water level, the rest of Gertie was never built and her body rests on a concrete block foundation, just like the *Empress Lilly* paddle wheeler at Disney Springs).

At the time, the area was known not as Echo Lake or Echo Park, but as Lakeside Circle, a term that went out of fashion. The image of the Gertie building was used constantly in publicity, including by *Time* magazine in the early days of the park where it became as much an icon as the mouse-eared water tower.

As the nearby plaque to the building states:

The themed style of the building is known as "California Crazy" architecture. It became popular in the 1930s and was designed to attract the attention of potential customers in a big way.

Formally, this is known as "programmatic architecture". Another example at Disney's Hollywood Studios is the Darkroom building on Hollywood Boulevard that looks like a giant camera, but sold photo supplies inside. It is based on an actual photo shop from 1940s Hollywood.

California Crazy architecture meant the building would be a visual representation of what was inside, like the famous Tail of the Pup hot dog stand built in 1946 in Los Angeles that was in the shape of a giant hot dog in a bun, or the Randy's Donuts shop with its massive doughnut on top of the building, also in Los Angeles.

Back in the 1930s and 1940s, the time frame represented by Hollywood Studios originally, people believed it was the Ice Age that killed off the dinosaurs. That's why it is the ice cream of "extinction" rather than "distinction" that is being sold at this location.

If you watch closely, Gertie is so cold that steam occasionally comes out of her nostrils. The top part of her is covered with snow. In her original concept sketch, and when she first appeared in the park, the green words "Ice Cream" covered with snow curved over the top of her back, but, over the years, that lettering was removed. It was definitely still there in 1992.

Gertie is in a lake because in her animated cartoon, she is by a large lake throughout. In the film she is white, but she is colored green at the park because the first movie posters of her were sometimes colored green, due to the popular belief at the time that dinosaurs were green or brown in color, like reptiles.

If you follow the pathway near her to a set of steps behind her, you will see on the walkway where Gertie's feet have cracked the cement as she walked into the lake and left an imprint. In addition, when the park first opened, those footprints continued in a floral format in the landscaped bed behind Gertie.

However, according to proposed plans for Disney's Hollywood Studios, it is time to say good-bye to dear Gertie.

For business reasons, and to meet the needs and wants of current guests, Disney parks always change. Beloved attractions and large acres of land are replaced with newer entertainments.

Currently on Disney's Imagineering drawing boards are plans for expanding the presence of the *Star Wars* series of films at Disney's Hollywood Studios in the area where Gertie now quietly sits observing the many guests who have no idea that she was once a superstar.

The Oswald the Lucky Rabbit Story

Oswald the Lucky Rabbit was not stolen from Walt Disney.

That's one of those popular "everybody knows this is true" Disney myths that continues to this very day.

Oswald couldn't have been stolen from Walt, because Walt did not own the character. Universal Pictures owned the character.

An animated cartoon series featuring a rabbit character had been suggested by the founder of Universal, Carl Laemmle. Walt knew that Universal owned the character. He was not shocked about that fact when he had a meeting with Charles Mintz in February 1928 to try to negotiate for a slightly higher budget for each cartoon. Walt was well aware that he was just a freelance contractor who had been hired to do the cartoons.

However, Walt was surprised that Mintz had signed Walt's animators to contracts and was going to set up his own animation studio in Hollywood run by Mintz's brother-in-law, George Winkler. This studio would produce any future Oswald the Rabbit cartoons.

Mintz offered Walt a high-paying job in the new studio. It was a job that Walt turned down because he would just be another Mintz employee with no control over the cartoons or the character.

Walt was also surprised that Mintz had successfully convinced Universal Pictures that the cartoons would not undergo a slip in quality as a result of the switchover, because they would still be made primarily by Walt's own staff. More importantly, Mintz had told Universal that the second season of cartoons could be made for roughly the same price as the first season, not the higher costs that Walt wanted.

While in New York for that fateful meeting, Walt had even met with Universal executives to try to get them to intercede on his behalf in his negotiations with Mintz. They were sympathetic to Walt's arguments, but ultimately felt that Mintz, who was older and had more business experience, knew what he was doing.

Over the years, Walt talked about how Oswald had been taken from him by the unscrupulous dealings of Mintz—and that is true. However,

Walt's statements have often been misinterpreted to mean that something illegal had been done, as a result of which Walt lost his personal property.

It was not unusual for a studio to own the rights to its characters. Animation legends Chuck Jones (director) and Michael Maltese (writer) created some iconic characters, like the Road Runner and Wile E. Coyote and the skunk Pepe Le Pew. However, those characters, as they knew, were owned completely by Warner Bros, as were all the other classic characters they created while working there.

Warners Bros could, and did, have others do cartoons with those characters. There was no legal obligation to credit or compensate Jones and Maltese in any way. It was only decades after they originally did the work that Jones and Maltese started to receive any public recognition at all.

Walt was just doing "work for hire" for Mintz, who had a deal with Universal to provide animated cartoon shorts featuring Oswald for the studio to distribute. But Walt felt that because of all the time and effort his studio had invested in the 26 cartoons to make Oswald a success, he was owed some greater consideration. Morally, that might have been the case, but there was nothing illegal in the proceedings.

As they still say today: "It's not personal; it's business."

Mintz took the opportunity to make this change just before a new contract was signed. In addition, Mintz was already in control of Ben Harrison's and Manny Gould's Krazy Kat production house in New York and enjoyed being in a position to control the character and the production costs.

As far as Mintz was concerned, Walt was a troublemaking maverick who didn't understand business realities and wasted time and money on unnecessary improvements. In addition, he knew that many of Disney's key staff (who were roughly the same age as their boss) disliked Walt's authoritarian manner. Mintz felt that Walt's contributions were valuable and had helped make the cartoons a success. However, he felt that Walt needed to be brought under control to maximize profits.

Mintz reasoned, as did so many others over the years, to their regret, that since Walt was not drawing the cartoons, he was not the real secret of the success of the product. It was the artists who were the key to the magic.

It is also important to remember that if Walt had had his contract renewed to keep making Oswald cartoons, there never would have been a Mickey Mouse. Oswald may have eventually developed into a character similar to Mickey, but it wouldn't have ever been quite the same.

Cartoon authority David Gerstein said:

> I always felt that the Mickey Mouse cartoon *The Barn Dance* (1929) could very easily have been an Oswald cartoon. Mickey is portrayed as an Oswald-like fall guy who—uncharacteristically for Mickey—loses in the end, to Pete no less!

It was Mickey's personality based on Walt himself that made him not only a star like Oswald (who appeared in 194 short cartoons), but a genuine cartoon superstar like the early Felix the Cat.

So, rather than reviling Mintz as a villainous businessman, we may all owe him a debt of gratitude. By not renewing Walt's contract, it forced Walt to come up with something better and allowed Walt autonomy over his studio and its films.

Walt Disney's animation career began in Kansas City, Missouri, where he produced a series of short, silent, black-and-white *Laugh-O-Gram* cartoons that took classic folk tales like *Little Red Riding Hood* and *Cinderella*, and updated them to a Jazz Age sensibility. Unfortunately, Walt's lack of business acumen resulted in the company going bankrupt after just seven silent cartoons being produced. Thankfully, all seven of those cartoons exist today, with the remaining three that were lost for decades finally being located in 2010.

In a desperate attempt to save his animation studio, Walt did several experiments including a "sing-a-long" short and an animated film that combined a live action girl with animated characters. That film, *Alice's Wonderland* (1923), caught the eye of film distributor Margaret Winkler when Walt sent it to her. Winkler had been the secretary of Harry Warner at the Warner Bros studio and with his encouragement became a cartoon film distributor.

Officially, she represented herself as "M.J. Winkler" so that people would think she was a man. That "J" stood for nothing. She just added it because she thought it looked better.

Winkler was barely 28 years old when she signed Walt Disney to produce the Alice Comedies. She was distributing two cartoon series: Pat Sullivan's Felix the Cat and Max Fleischer's Out of the Inkwell. Fleischer was going to be setting up his own studio and the mercurial Sullivan was currently unhappy with Winkler and had some outrageous demands for a new contract.

So, Winkler was on the verge of losing her only two cartoon series when she contacted Walt. That's one of the reasons the Alice Comedies feature a black cat (Julius). Winkler felt his role could be expanded if Sullivan did not renew his contract and took Felix the Cat away.

Sullivan eventually did renew the contract in 1924, but opted out in 1925. As David Gerstein notes, Winkler indeed demanded Disney expand Julius' role at the time. Winkler also launched a new licensed series built around Krazy Kat, another popular black cat character she could utilize in Felix's absence.

The Disney Bros. Studio was formed in 1923 to produce a series of silent Alice Comedies cartoons featuring little six-year-old Virginia Davis interacting with cartoon characters and backgrounds. That series is the true beginning of the Disney animation empire. Fifty-seven cartoons in the series were produced, and it was so successful that both Walt and Roy could afford to get married and build houses for themselves.

In 1924, Winkler married Mintz (who had been working for her since 1922), had a baby, and retired from the business. Her new husband took over and his hardheaded business approach resulted in several confrontations with the Disneys.

Where Winkler had been supportive of Walt and Roy, Mintz was highly critical that the cartoons were not of a high enough standard for him. In later years, Walt admitted in an interview that his desire for high standards was cultivated through his interactions with Mintz.

The last Alice Comedies featured less and less live action, and more and more animation as Walt's animators gained in expertise. Walt felt the "gimmick" of live action/animation in the series had run its course and was losing favor with the audience. However, because of the popularity of Walt's cartoons, in 1927 Mintz was able to sign a deal with Universal to produce a new cartoon series, which was to star a rabbit character.

It was Carl Laemmle himself, the founder of Universal, who said that there were too many darn cats in cartoons, like Felix and Krazy. He wanted to see a rabbit.

The name "Oswald" was reportedly selected by P.D. Cochrane, head of Universal's publicity department. He gathered suggestions from the staff around the office, including the secretaries, and put them all into a hat and drew out a name.

Walt would later tell his daughter Diane how the name was literally drawn out of a hat at random which is why Oswald didn't follow the animation and comic strip convention of alliteration in a name, like Mickey Mouse, Bugs Bunny, Pink Panther, and others.

Universal had not released a cartoon series in years, and began to promote Oswald with extensive and enthusiastic advertisements in the trade press, although the odd-looking white rabbit in their earliest ads had not the slightest resemblance to the final Ub Iwerks black rabbit design.

Universal did a marketing push on merchandise, as well, including a 5-cent chocolate-covered marshmallow candy bar made by the Vogan Candy Corporation of Portland, Oregon. The Philadelphia Badge Company issued a button with Oswald (the precursor to today's pin collecting). Universal Tag and Novelty Company offered an Oswald stencil set for drawing the character.

The Disney Studio saw no royalties from any of this merchandise, which didn't bother Roy, who said, "We are a movie studio, not a toy store." The Disneys were happy that the merchandise brought attention to the character so that audiences would ask theaters when they would be showing another Oswald cartoon. In any case, merchandise profits would have gone to Universal who owned the character and the cartoons.

Winkler suggested to her husband that Walt and Roy should be ones to make the series. Mintz offered Disney the contract to produce the new series, and Walt jumped at the opportunity. Almost every artist at the studio submitted different rabbit designs.

Iwerks had been experimenting with a new way of designing characters for animation using circles, which were easier to animate. If you look closely at the last few Alice Comedies, the ears of the mice got longer and more pointed as Iwerks practiced with how to move rabbit ear shapes.

Disney sent Mintz the group of test drawings for the new character, and Mintz selected and approved one. It was copyrighted for Universal. Disney was given approval to produce the first cartoon for the new character, who was to be known officially as Oswald the Lucky Rabbit.

The pilot, titled *Poor Papa*, was criticized by Universal executives because the Oswald character was not cute or likeable enough. In fact, Gerstein has described the character as looking and behaving like "Farmer Alfalfa" from the Paul Terry cartoons with the same type of grumpiness and dumpiness of a stereotypical rural hick.

The memo from Universal included the phrase: "With the exception of Chaplin, important movie comedians are neat and dapper chaps." The Oswald that Iwerks had originally designed had overalls, jowls, and a scruffy shape in keeping with the plot of the cartoon.

In the story, Oswald was overwhelmed with a multitude of newborn children. The concept, Gerstein told me, was decided upon before the character's species was nailed down, but it was ideal for a rabbit character, given the belief that rabbits had a tendency to multiply abundantly and quickly. This was a storyline Walt would refine for a later Mickey Mouse cartoon titled *Mickey's Nightmare* (1932).

After viewing *Poor Papa*, Mintz advised Disney to redesign the character so that he was "young and snappy looking with a monocle".

While Universal had declared that *Poor Papa* was "unrelease-able", when Mintz later took over the series he was forced to release it in an attempt to keep up with the deadline schedule set by Universal.

Iwerks ignored Mintz's suggestion for adding a monocle, but refined Oswald's appearance so that he appeared younger and more vital. In fact, this new design had many of the same design elements that we associate with the early Mickey Mouse. As the series progressed, Oswald became more and more the prototype for Mickey. However, Oswald was always a little "rougher" in behavior and more anatomically flexible, able to pop off parts of his body like Felix the Cat and utilize them in a variety of ways.

The second Oswald cartoon, *Trolley Troubles*, was acceptable, and was previewed on July 4, 1927 to terrific reviews in the trade press. Its official premiere date was September 5, at which point it went into wide release. Motion Picture World gushed that: "Oswald series has accomplished the astounding feat of jumping into the first-run favor overnight."

Trolley Troubles was inspired by Fontaine Fox's popular comic strip *Toonerville Trolley*. Oswald is the long-time conductor on this hometown trolley car that becomes a runaway threat careening up and down hills. There is a scene where Oswald removes his foot and rubs himself for luck. Friz Freleng animated it, and much later groused in an interview:

> And I was questioning, "What do I show when his foot's taken off, do I show a bone in there or what?" And Walt joked about it and of course, he never thought of it either. Nobody had thought of it.

Why did Oswald rub himself with his foot? It never seems to occur to anyone that Oswald is a "lucky" rabbit because he has four lucky rabbit feet.

Carrying a rabbit foot was a lucky charm for folks in those days. The custom lasted through most of the 20th century when the novelty feet were sometimes colored in a variety of shades like green or orange to increase sales. And, yes, you could actually feel the foot and the claws underneath the hair.

The level of craftsmanship and storytelling continued to improve in the cartoons which, as a result, began to increase in cost. The animators began to rely less and less on model sheets as a reference for tracing poses. Accordingly, the animation began to have a more fluid style. Another change was that, although the storyboard method was still years in the future, the shorts began to be scripted with story sketches to indicate posing and gags.

Several gags in the cartoons were later re-done in early Mickey Mouse cartoons. In fact, Walt's script for *Steamboat Willie* references a gag from the Oswald cartoon *Tall Timber*. Even Peg-Leg Pete popped up occasionally as a villain for Oswald as he had earlier menaced Alice in the Alice Comedies.

Walt divided his animation staff into two separate units so that two pictures could be produced at once. One unit included Ub Iwerks and Friz Freleng. The other was headed by Hugh Harman and Ham Hamilton.

The Oswald cartoons ran at the prestigious Colony Theater in New York. It was the same theater that would showcase the debut of Mickey Mouse because the owner was familiar with the quality of Disney's Oswald cartoons.

In 1928, with the popularity of Oswald growing, Disney went to New York to approach his distributor, Mintz, about an increase in his budget. Walt hoped the increase would allow for more improvements. Mintz not only refused, but told Walt to accept a 20 percent cut in the budget, or else

Universal, the legal owner of the series, would produce it on its own using Walt's own animators, who had signed contracts with Mintz.

Legend states that only the loyal Iwerks refused to sign a contract with Mintz. In actuality, Les Clark (later to be known as the "Mickey expert" and one of the fabled Nine Old Men) and Johnny Cannon also refused to sign.

So that left Walt with only a staff of three animators, two of whom who might be considered just apprentices since they had limited experience in animation.

Disney refused to sign the contract and formally walked away in March 1928, boarding a train to Los Angeles after sending his brother Roy a telegram assuring him that everything was all right.

To fulfill his current contract, Walt still had to make three more Oswald cartoons with the "renegade" animators who would be leaving in June for the new studio. Walt tried to push those through production as quickly as possible, while Iwerks was separated and hidden from the rest of the staff so he could work on the first Mickey Mouse cartoon.

Mintz gave the series to his brother-in-law, George Winkler, to produce in a new studio. Winkler Studio was renamed Mintz Studio in 1929. The first Oswald cartoon produced by Winkler was *High Up*, released on July 23, 1928, roughly four months before the debut of Mickey Mouse. Mintz decided he could make the cartoons at his own studio with Walt's animators, whom he had hired away from an unsuspecting Walt.

When Walt's contract for making Oswald cartoons was not renewed, supposedly Walt confronted Charles Mintz and said, "Protect yourself, Charlie. If my artists did this to me, they'll do it to you." Mintz laughed and didn't believe him.

The Winkler studio lost the contract to produce more Oswald cartoons in 1929 when animators Hugh Harman and Rudy Ising went to Universal to try and convince the studio to put them in charge instead of Mintz. Carl Laemmle, who was the head of Universal Pictures, was tired of all these internal politics. He decided to produce the series in-house with director Walter Lantz taking charge.

Lantz had been working in animation in New York since 1922 before moving to Hollywood. At the Winkler studio, he supplied gags and became a director on the Oswald series starting with *Mississippi Mud* (1928). Pinto Colvig (later a Disney storyman and the voice of Goofy) also supplied gags for Oswald.

More importantly, Lantz played poker with Laemmle every Thursday night, and Laemmle considered him a lucky charm. "He's been lucky for me at poker," Laemmle said, "so maybe he will be lucky for me at producing cartoons."

Lantz had a drink with Walt Disney to see if this situation would cause Walt any concern. Walt, who was now successful with Mickey Mouse, gave Lantz his blessing and told him there would be no hard feelings. They remained friendly for the rest of their lives.

As Walt told his daughter Diane in the summer of 1956:

> An old-timer in the business slipped in before they [Walt's former animators] could put over their fast move and took charge of the Rabbit. I was cheering for him throughout all of that infighting, and I got a kick out of it when he outsmarted the artists who'd deserted me.

Mintz was out in the cold and Harman and Ising found work at Warner Bros creating the first Looney Tunes and Merrie Melodies cartoons.

Lantz, working with animation legend Bill Nolan (who was just as fast and talented as Iwerks), slowly turned Oswald into a cuter, more childlike character. Oswald got white gloves, shoes, and larger eyes.

Another design change came in 1936 when Lantz assigned Manuel Moreno to redesign the character into a white-furred, chubbier rabbit. This new design didn't find favor with audiences. In addition, it severed any connection at all with Walt's version of the rabbit.

Lantz's release schedule included several reissues of some of the original Disney cartoons, now with added soundtracks. Sound had been added to the series midway through the Winkler season. In the Lantz era, a variety of people, including Mickey Rooney, June Foray, and Lantz himself, supplied Oswald's voice.

The series wore out its welcome and ended in 1938 (with one additional cartoon, *The Egg Cracker Suite*, in 1943), although Oswald still appeared for many years in the Dell comic book series featuring Lantz's characters.

Starting in the late 1940s comic books, Oswald was portrayed as a brown adult rabbit who adopted two young orphan rabbits, Floyd and Lloyd. This version of the character had absolutely nothing in common with the earlier Oswald from the Disney, Winkler, and even early Lantz cartoons, but it lasted in comic books for roughly two decades.

Lantz developed Andy Panda in 1939 and Woody Woodpecker in 1940 who became even more successful and popular than Oswald ever was, which was one of the reasons Oswald went into retirement.

There is an urban legend that Walter Lantz won the ownership of Oswald the Lucky Rabbit from Carl Laemmle in a poker game in the 1930s. The actual truth is that when Laemmle was forced out of Universal in 1936, Lantz was clever enough to see that Universal would probably eliminate the cartoon studio, so he renegotiated his contract.

He became an independent producer supplying Universal with animated shorts, and took control of the copyrights and trademarks for all of the characters he had worked on, including Pooch the Pup and Oswald the Rabbit. Universal seems to have felt it had become an unnecessary burden to manage the characters and the previously made cartoons, and thought it was a good business to have cartoons produced by an outside contractor, eliminating overhead at the studio.

So, for decades, Lantz was the owner of Oswald. In 1984, he sold everything back to MCA/Universal, but remained active until his death consulting on how his characters would be used in theme parks, comic books, merchandise, video, and other venues.

In February 2006, Disney CEO Bob Iger agreed to a trade with NBC Universal for a number of minor assets in return for releasing sportscaster Al Michaels from his ABC and ESPN employment contract to go to NBC Sports.

What made the real news is that one of those minor assets was the rights to the Oswald the Lucky Rabbit character designed by Disney, and the original 26 short cartoons made by Walt. All the rights to the other Oswald cartoons and merchandise produced by Winkler and Lantz were retained by Universal.

In early 2006, as permitted by the deal with Universal, Disney filed for numerous Oswald-related trademarks to secure the Disney version of the character. The later versions of the character had trademarks that linked them to Lantz and Universal.

The media exploited the fact that "Oswald comes home", and in a statement made on March 9, 2006, Diane Disney Miller said:

> When Bob [Disney president and CEO Robert Iger] was named CEO, he told me he wanted to bring Oswald back to Disney, and I appreciate that he is a man of his word. Having Oswald around again is going to be a lot of fun.

In December 2007, thanks to the efforts of film historian Leonard Maltin, animation historian David Gerstein, and a Disney restoration team that included Theo Gluck and Steve Poehlein, 13 of the existing Disney Oswalds were released on DVD as part of the Walt Disney Treasures series.

In addition, Disney also released a new line of character merchandise and included Oswald as a major character in the 2010 Mickey Mouse video game *Epic Mickey* as well as two follow-up games, *Epic Mickey 2: The Power of Two* and *Epic Mickey: Power of Illusion*.

The full version of the Oswald cartoon *Oh, What a Knight* is included as an unlockable cartoon in *Epic Mickey* by collecting film reels in the game.

Just like Duffy the Bear, Oswald the Rabbit is hugely popular in Japan with Disney offering key rings, puppets, inflatable dolls, clothing items and other assorted merchandise to the eager fans.

With the re-opening of Disney California Adventure in 2012, at the front of the park is Oswald's Service Station, a gift shop themed like a 1920s gas station. A plethora of Oswald merchandise, including a rabbit ears cap reminiscent of the famous Disney theme park mouse ears cap, decorates the area.

Tokyo DisneySea introduced a new walkaround costumed character version of Oswald in April 2014 and a version of that costume began making appearances at Disney California Adventure on September 14.

As Disney fans, we should be grateful that any of the Disney Oswalds exist at all. Of the 26 Oswald cartoons made by the Winkler-Mintz studio, only about a dozen seem to have survived. Fortunately, more of the Disney-produced Oswalds survive. A couple more have been recently located and are in the process of being restored.

The fact that any silent cartoons at all exists is something of a miracle.

One of the reasons that some of these silent Oswald cartoons still even exist is that, as noted earlier, Walter Lantz reissued several of the early Disney Oswald cartoon with added sound. In 1931, due to budget restraints and schedule challenges, these re-releases were necessary to fill gaps in the production schedule. James Dietrich did the soundtracks.

Other Disney Oswalds came to the home movie scene by different means. *Oswald und die Wolkenkratzer*, a localized German print of *The Sky Scrapper* (1928), was translated back to English as simply *Skyscraper*, and sold for many years in an unauthorized 16mm release. Today, *Skyscraper* still exists only because a dedicated researcher located two incomplete but different prints of the cartoon and combined them to recreate the original.

As researched by Gerstein, these Oswald reissues were later used by a company called Motion Pictures for Television (MPFT) during the early 1950s. On its own and through U.M.&M. TV Corp, MPFT distributed a large selection of black-and-white Lantz sound shorts to television and home movie markets.

This Disney Oswald cartoon *Skyscraper* greatly influenced the storyline and gags in the popular Mickey Mouse short *Building a Building* (1933).

As Gerstein explained when we talked at length about the character and the series:

> The decades-long absence of an original English language print led *The Sky Scrapper* to be inaccurately listed for those decades under that working title including on the 2007 DVD.

More recent research has led to the original title's proper identification. This is yet another example of why Disney history is so exciting in recent years with new discoveries resurfacing in surprising ways, including films that were misidentified.

In a strange way, many of the Disney Oswald cartoons survived when those who worked on those cartoons remade them or used major elements from them with other cartoon stars. Oswald's *Harem Scarem* (1927) became Disney's later *Mickey in Arabia* (1932). Oswald's *Rival Romeos* (1928) became Ub Iwerks' Flip the Frog short *Ragtime Romeo* (1931).

When Mickey Mouse tipped his ears like a hat in *The Karnival Kid* (1929), inspiring the mouse-eared caps in Disney theme parks, the gag had already been done in the Oswald cartoon *Sleigh Bells* (1928) nearly a year earlier.

In *Steamboat Willie* (1928), when Pete pulls Mickey's stomach and it stretches, Walt's comment on the original story sketches outline was "same as Oswald and the Bear in *Tall Timber* (1928)". When the goat eats the sheet music and Mickey cranks its tail for the music to play, Oswald had already done it in *Rival Romeos* released almost nine months earlier.

According to a Library of Congress survey, more than 85% of silent movies produced in the United States have already been lost or are in unrestorable condition. Another survey claims that more than 90 percent of silent movies produced before 1930 have been lost.

With each passing year that percentage increases, despite the efforts of various people.

Silent cartoons were produced in small quantities and then those prints were circulated to theaters in the United States and then later shipped overseas until the prints often just fell apart or were never returned.

When the "talkies" era came into vogue, silent films were considered worthless, and then later when color became standard in films, the black and white films became even less valuable.

During the silent era, cellulose nitrate film was used for the majority of films. It is a highly flammable and unstable compound, with a life span of between 30–80 years.

The historic *Gertie the Dinosaur* survives in a complete form only because there were multiple copies of it found in 1947 in a garage of a friend of Winsor McCay who had made the film and then literally threw away the copies as he moved on to other things and needed storage space.

The cans containing the film (and other films by McCay) were opened in a barrel of water so the film wouldn't instantly burst into flame. The decomposition of nitrate film cannot be halted, although in the right conditions, it can be slowed so that a copy can be made.

Cellulose nitrate was first used as a base for photographic roll film by George Eastman in 1889 and was used for photographic and professional 35mm motion picture film until 1951. The silver nitrate gives the images a sharpness and warmth that resulted in the term "silver screen" being coined.

It is highly flammable and also decomposes to a dangerous condition with age. When new, nitrate film can be ignited with the heat of a cigarette. Nitrate film burns rapidly, fueled by its own oxygen, and releases toxic fumes.

Many films were lost in studio fires caused by this decomposition. Some films were destroyed deliberately for their silver content, while others were just allowed to decompose due to simple neglect and lack of interest.

Laemmle, when he needed a bonfire for a scene in one of his "talkies", told his assistants to pull out some of the silent films that were loaded with silver nitrate and toss them onto the fire so it would glow brighter.

Universal, which distributed the Oswald cartoons, dumped its entire collection of remaining silent films in 1948 to free up storage space for its new films, and all of Samuel Goldwyn's earliest productions were supposedly destroyed to save money on insurance premiums and storage costs.

There are other reasons why films disappear. When Walt Disney decided to make the live-action feature *Swiss Family Robinson,* he bought up the rights to the 1940 version produced by RKO and confiscated all known prints so there wouldn't be comparisons to his remake.

This used to be standard operating procedure at all the major studios, and accounts for many missing films, which is one of the reasons the silent feature film version of *Peter Pan* disappeared from public view for decades.

Although the nitrate negative of the 1940 original *Swiss Family Robinson* was destroyed in a fire years ago, the ever-incredible Scott MacQueen, who was at that time the Disney film archivist, was able to get an original 16mm copy from a private collector, and also located a 35mm print he found in another archive and was able to piece together a complete, fairly good looking master copy of the film.

Oswald the Lucky Rabbit has had a long and colorful life, surviving long after many of his contemporaries became vague memories. The Disney version of the character is especially appealing.

Here is a complete list of the Disney-produced Oswald the Rabbit cartoons. Disney made 26 Oswald cartoons. (Winkler made another 26 and Lantz another 142 of them.)

1927

- Trolley Troubles (September 5, 1927)
- Oh, Teacher (September 19, 1927)
- Great Guns (October 17, 1927)
- The Mechanical Cow (October 3, 1927)
- All Wet (October 31, 1927)
- The Ocean Hop (November 14, 1927)

- The Banker's Daughter (November 28, 1927)
- Empty Socks (December 12, 1927)
- Rickety Gin (December 26, 1927)

1928

- Harem Scarem (January 9, 1928)
- Neck 'n' Neck (January 23, 1928)
- The Ol' Swimmin' 'Ole (February 6, 1928)
- Africa Before Dark (February 20, 1928)
- Rival Romeos (March 5, 1928)
- Bright Lights (March 19, 1928)
- Sagebrush Sadie (April 2, 1928)
- Ride'em Plow Boy (April 16, 1928)
- Sky Scrappers (April 1928)
- Ozzie of the Mounted (April 30, 1928)
- Hungry Hoboes (May 14, 1928)
- Oh, What a Knight (May/June 1928)
- The Sky Scrapper (June 11, 1928)
- The Fox Chase (June 25, 1928)
- Tall Timber (July 9, 1928)
- Sleigh Bells (July 23, 1928)
- Poor Papa (August 6, 1928)
- Hot Dogs (August 20, 1928)

PART THREE
Disney Park Stories

The Disney theme parks are the most popular vacation destinations in the world. Many people have made annual pilgrimages to them for decades. They are universally considered to be unique and high quality experiences. In 2015, nine of the top ten amusement parks in the entire world were Disney parks.

(The non-Disney park in the Top Ten? Universal Studios Japan in Osaka.)

Yet, in 1953, Walt Disney Productions stockholders did not agree that the company should be getting involved in such a project. When Walt was prepared to buy property in a beautiful valley in Calabasas, California, for the project, the stockholders filed and won a lawsuit to prevent him from doing so.

In an impassioned and tearful address to the stockholders, Walt stated:

> I don't want this company to stand still. We have prospered before when we have taken chances and tried new things. This is our golden opportunity—a chance to move into an entirely new field.
>
> You say we are not in the amusement park business. No, we're not. But we are in the entertainment business. And amusement parks are entertainment.
>
> I know it is difficult for you to envision Disneyland the way I can. This kind of thing has never been done before. There's nothing like it in the entire world. I know, because I've looked. That's why it can be great because it will be unique. A new concept in entertainment, and I think...I KNOW...it can be a success!

Even Walt's older brother Roy was not thrilled with the idea and limited the studio's initial investment to just $10,000.

Walt raised $100,000 by borrowing on his life insurance policy and then also sold his home in Palm Springs to finance his own separate design company, WED Enterprises (now known as Walt Disney Imagineering). Walt took one of the biggest risks in his career and it is still paying off today.

On the morning of August 8, 1953, Walt reviewed the latest of the dozens of Disneyland site maps that Imagineer Marvin Davis had worked on and picked up a No. 1 carbon pencil and drew a triangle around the plot of land to indicate where he wanted his railroad to run.

"I just want it to look like nothing else in the world," said Walt. "And it should be surrounded by a train."

There are many reasons for the Disney parks to be considered so different and popular. People often cite the lush landscaping or the attention to detail or cleanliness or even the emphasis on safety. However, at the top of all of the lists is storytelling. The Disney parks were the first amusement venues to consistently tell a coherent story, not only in the individual attractions, but in the entire design. That's why they are theme parks.

Attractions had a definite beginning, middle, and an end. They did not necessarily recount a chronological retelling of a story, but their use of familiar scenes and images helped guests remember the story and become part of the overall experience. Everything that was included was used to evoke the same story and not contradict it.

For many including myself, the most fascinating stories are the stories behind the attractions themselves and what choices were made on how to tell the stories the guests would experience.

This section documents some of those stories that were never shared with guests and not even cast members in some cases, but have added significantly to the magic that makes the Disney parks so different.

Walt Disney World's Tomorrowland

The Tomorrowland Light and Power Company in Walt Disney World's Magic Kingdom closed permanently on February 9, 2015. The 15,000 square-foot space at the exit of Space Mountain was filled with arcade games and merchandise, and was designed by McMillen Design International.

While Disney made no formal announcement, other than the boiler plate statement "we are always evaluating our offerings and making adjustments", it likely reacted to changes in Florida law regarding internet cafés (with the law stating, among other things, that a person can not win prizes worth more than seventy-five cents and that games need to be coin-operated). Rather than face a potential legal battle down the line, should interpretation of the law broaden, Disney shut down the Tomorrowland arcade.

Most Disney fans probably never used the term "Tomorrowland Light and Power Company", calling it "the video arcade near the exit of Space Mountain" instead. The arcade was part of a larger branding effort of the entire land by Imagineering in 1994 to create "the future that never was.... is finally here!"

According to the backstory, a metropolis the size of the Tomorrowland community, which is the home of the League of Planets, needs substantial power to operate, and so a proper power facility like the Tomorrowland Light and Power Company is vital. The building that houses this facility has its own distinctive logo and futuristic power transformers adorn its exterior.

But where does the power come from?

A row of mechanical metal palm trees dotting the area in front of Space Mountain are, in fact, "Power Palms". When their metallic fronds are extended, they capture solar energy and store it in the round, metallic, coconut-like globes that surround the top of the trunk. As they collect this energy, the globes glow, lightning the area at night as a byproduct.

One Power Palm is purposely stuck in a half-way position with its fronds drooping and its globes missing because they were supposedly harvested to remove the energy. Once that task has been accomplished, the globes will be returned to gather even more solar power until they are once again full, and the process will repeat itself.

By the way, with Tomorrowland's renovation, Space Mountain got its first official future name: Starport Seven-Five, a reference to the official January 1975 opening date of the attraction at Walt Disney World. After all, the term Space Mountain sounds like an attraction rather than a functioning building you would find in a city.

And it is all just part of a much larger story. However, this vision was never really shared with guests and cast members. It was not even part of the official training required to work in this area.

Walt Disney believed that the stories in the parks needed to be apparent and be shared with both guests and cast so that they could support it and enhance it. As a result, standard operating procedures booklets at Disneyland originally included the history and storyline of the attraction, whether it was the Submarine Voyage (with commentary from Admiral Joe Fowler) or Great Moments With Mr. Lincoln.

A story is a living thing, and if you don't take care of it, then it dies. Things get added or removed from the story because of this unawareness and the story fades or mutates into something unrecognizable or inaccurate.

That's the story of Walt Disney World's Tomorrowland today. What was once a clever, consistent storyline that held together the theme of the entire land has been so forgotten that the closing of an obscure building brings no concerns of how that impacts the rest of the story.

It was Walt's intent that Disneyland's Tomorrowland would reflect a centralized spaceport of "the future that was just around the corner" and showcase the latest in technology that would make "a great big beautiful tomorrow just a dream away" for everyone.

In 1955, Disneyland's Tomorrowland featured no outrageous flying saucers, atomized rayguns, or little green men from Mars. Instead, it offered a simulated flight to the moon based on authentic scientific information supplied by scientists like Werhner Von Braun and Willie Ley.

Great care was taken not only to accurately show the known landscape of the moon, but what its never-seen dark side might look like, after carefully consulting with experts at the Griffith Observatory.

Perhaps because of that, Tomorrowland proved to be Walt's most frustrating challenge. "The only problem with anything of tomorrow is that at the pace we're going right now, tomorrow would catch up with us before we get it built," Walt reportedly said. During his lifetime, he made two major overhauls of the land, in 1959 and 1967, and still was unsatisfied.

When Disneyland Paris (Euro Disneyland) opened in April 1992, it did not have the traditional Tomorrowland, but a substitute called

Discoveryland. It was purposely themed to "yesterday's future" with an emphasis on the 19th-century-style future envisioned by French writer Jules Verne and the latest scientific discoveries. In this way, the land would not need to be continually updated like a Tomorrowland, but have a sense of timelessness and fantasy.

So when it came time to update Walt Disney World's Tomorrowland just two years later, in 1994, the Imagineers decided to Americanize that concept by re-making it into "the future that never was".

That is to say, it would be the future that resembled the one predicted in all the science fiction magazines and movies of the early 20th century. It would be a future that would never be out of date and also be able to incorporate some humor that was sorely lacking in earlier versions.

Imagineer Alex Wright said:

> New Tomorrowland [of 1994] was conceived as the meeting place of the universe. It is an interplanetary hub chosen to serve as the head-quarters of the League of Planets. Everything in this land relates to excitement and optimism about the future. Every detail relates to this theme.
>
> Ours is a retro-future concept replete with all the trappings of an intergalactic spaceport. We all remember when we thought the future would be like this. Tomorrowland offers us the opportunity to visit it.

In keeping with the theme that Tomorrowland is a city that exists in some alternative version of the future, at the entrance is a huge sign from the Tomorrowland Chamber of Commerce that welcomes guests with its motto: "The Future That Never Was Is Finally Here".

This is a metropolis where humans intermingle with aliens and robots and is the headquarters for the League of Planets, an organization that governs the peaceful universe with a firm but just hand.

Just like Main Street, U.S.A., there is also a main street in Tomorrowland— the Avenue of the Planets. And, as in many cities, there is a Chamber of Commerce posting at the entrance, but without the shields of the more familiar Lions Club or Kiwanis organizations. Instead, there are colorful emblem logos representing organizations like The League of Planets, The Loyal Order of Little Green Beings, Galactic Association of Retired Aliens (rather than the AARP), and the Sleepless Knights of the Milky Way (that's "Knights" with a "K", bringing to mind the fraternal Knights of Columbus, as well as the mighty heroic warriors that sometimes filled the pages of science fiction pulp magazines of the 1930s). In addition, it indicates that a person can't sleep at night in the Milky Way because all those darn stars are shining so brightly all the time.

On the left side of the entrance, there was the InterPlanetary Convention Center with additional advertising posters inside about upcoming shows and exhibits. In 1995, the building had a demonstration of the latest in teleportation technology from X-S Tech. In 2004, it was converted into the Galactic Federation Prisoner Teleportation Center to handle "undesirables". Residents and tourists can see a demonstration (just like in a regular court house) of how justice is done. In this case, convicted prisoners are teleported away from this golden city of tomorrow rather than sent to the local prison. In fact, today, Experiment 626 (commonly known as Stitch) is to be dispatched.

On the right side of the entrance was the Tomorrowland Metropolis Science Center, featuring a demonstration by The Timekeeper about time travel. Today, that building is the home of the Tomorrowland Expo Center (taking on the role of the former convention center that closed across the street) hosting The Monsters Inc. Laugh Floor comedy club that opened in April 2007.

The exteriors of the buildings sport clever and amusing posters for the Tomorrowland Towers Hover Hotel (three miles directly above Tomorrowland on the Atmosphere Three Skyway Exit); Leonard Burnedstar, who will be conducting the Martian Pops Orchestra; the Space Home and Garden Show; and the Antique Rocket Show and Swoop Meet (featuring the fabled Moonliner from Disneyland's past). These posters indicate that there is a much larger community than guests will be able to experience, just as Center Street on Main Street, U.S.A. used to tell the story of "more Main Street just around the corner" before the Emporium extension was built.

Walking between these two buildings down the Avenue of the Planets, guests find themselves in the central hub of Rockettower Plaza. The names are a playful reference to New York's famous Rockefeller Plaza and the Avenue of the Americas.

It is at this hub where the main transportation system for the community is located, the Tomorrowland Transit Authority that was renamed the Tomorrowland Transit Authority PeopleMover in 2010 as an homage to its original 1975 title of the WEDWay PeopleMover. This attraction serves as an urban mass transit system for the citizens and visitors to this popular space port.

The Blue Line, which all guests and tourists seem to use, provides intra-city service to destinations throughout the city, from a beauty parlor to a merchandise shop. It is also the delivery method for businesses like Earth Crust Pizza. However, a close look at the signage reveals that there is also the Green Line for commuting to the Hoverburbs (the city's suburbs where people live) high above the city. In addition, there is the Red Line that takes riders off the planet to other destinations in the galaxy.

In 1994, the iconic Star Jets attraction was updated and renamed Astro Orbiter. The huge central rocket was replaced by a highly stylized iron-work launch tower along with various spinning planets on the outside of the attraction so that it seems like the rockets are weaving between the planets.

The new storyline is that the League of Planets was giving inexperienced pilots an opportunity to learn how to fly their own rocket ships before unleashing them into the universe. It is a sort of futuristic driver's training program.

Nearby is an actual working Metrophone booth from the Galactic Communications Network (GCN). As it states on the phone: "Bringing the World Closer Together. Toll Free from Anywhere in the Galaxy."

Since 1999, punching several numbers will bring up one of nine possible hilarious one-sided conversations from Rocket Realty, Sonny Eclipse's agent Johnny Jupiter, Earth Crust Pizza (delivering anywhere in the Solar System in less than two light years or your order is free), Intergalactic Movie Line (with information on the film *Attack of the 50 Foot Earthling*), Psychic Robots Network, and more.

Near the entrance to the TTA PeopleMover is a robot newsboy selling his newspapers. The Robo-Newz vendor is always up-to-minute and supposedly guests can get their daily paper printed "while u wait". The main case shows that the latest physical newspaper is a copy of the *Galaxy Gazette* with the headline: "Stitch Escapes!"

Those who wonder *where* Stitch escapes to can find the answer when leaving the Galactic Federation Prisoner Teleportation Center, otherwise known as Stitch's Great Escape. The upper part of his body is sticking out of the top of the ceiling in the Merchant of Venus (a humorous pun on Shakespeare's *Merchant of Venice*) merchandise shop. Unfortunately, the store uses the same unreliable teleportation technology as the prisoner teleportation center next door. It is laced throughout with blue cable coils from the ceiling that connect to collection orbs atop the fixtures so quality goods from throughout the universe can materialize in these mini-teleport chambers.

Of course, many other businesses operate in this community of tomorrow, including Mickey's Star Traders, a prominent importer and exporter of goods that does not use teleportation technology to stock its shelves. From the design inside the shop, it is apparent that the owners do indeed purchase their wares through a variety of sources, but that they are brought in by spacecraft. The artwork captures the 1930/1940 comic book art style of what the future was going to look like.

In the future, Coca-Cola maintains its predominance in the soft drink industry. The Thirst Rangers ("Delivering Refreshment to a Thirsty Galaxy") in their red-and-white (the traditional colors of Coke) rocket ship are perched high on a landing platform. At the bottom of the platform are gray crates (for standard transgalactic delivery) with images of a Coca-Cola bottle and amusing shipping labels and alien languages.

The Disney Imagineers created this spaceship out of the hull of the prop of the Trimaxian Drone Ship from the 1986 Disney film, *Flight of the Navigator*. One of those ships was originally on view for several years in the boneyard on the backstage tour at Disney-MGM Studios.

A good example of not taking the future too seriously is Cosmic Ray's Starlight Café, a popular chain of intergalactic franchise Starlight Cafés throughout the universe. "This is the FIRST Earth Restaurant Franchise from Outer Space!" proclaimed the original poster.

For a "limited time", guests dining in the Starlight Lounge can enjoy the song stylings and snappy banter of Audio-Animatronics performer Sonny Eclipse during his approximately twenty-plus-minute performance. Sonny was actually modified from a similar figure, Officer Zzzzyxxx, who was at the baggage screening desk outside of the Star Tours attraction at Tokyo Disneyland.

Direct from Yew Nork on the planet Zork, Sonny Eclipse is the "Biggest Little Star in the Galaxy". The Bossa Supernova and Eclipso musical stylings of Sonny and his Astro Organ, along with his ethereal and invisible backup singers The Space Angels, have entertained guests for almost two decades.

It's easy to get lost in the future, so there is a huge black globe at the entrance to the plaza near The Merchant of Venus that is a map to the universe. It weighs approximately six tons (more than 13,227 pounds) and rests on a fountain of water that allows it to rotate smoothly with the slightest touch.

This granite kugel ball shows all the routes you can take to get around, including Route 88, 44, and 5, but most importantly, Route 66. There are other symbols on this massive sphere, including where the nearest gas stations are to fill up your rocketship, and the nearest Metrophone booth. A careful eye will be able to detect Cinderella Castle on it, as well.

The live-action feature film *Tomorrowland* (2015) had no direct reference to this storyline, other than some whimsical touches like an image of Space Mountain in its skyline.

The closure of The Tomorrowland Light and Power Company did not spark any outcries from Disney fans, unlike such previous closures as Mr. Toad's Wild Ride, because no one knew its backstory.

Walt Disney World's Tomorrowland is not just the Future That Never Was, but the Future That No One Ever Knew.

The Story of Body Wars

When the Wonders of Life Pavilion at Epcot's Future World closed on January 1, 2007, there were Disney fans who did not shed a tear for the disappearance of a motion-control simulator known as Body Wars, a key attraction in the pavilion, described in the 1989 Epcot guidebook as:

> Science fiction and science fact merge with state-of-the-art simulator technology to propel you on a thrilling ride through the human body. Health and other boarding restrictions.

When it opened in October 1989, it was the first Epcot attraction to have restrictions on who could ride: guests had to be at least three years old and at least 40 inches tall.

The story behind Body Wars actually begins with Walt Disney himself.

Walt had a great fascination with miniatures. When the Monsanto House of the Future faced its final year at Disneyland, he was intrigued when Dr. Charles Allan Thomas, a key researcher at the Monsanto Chemical Company, approached him with the idea of creating an attraction that would explore the miniature world of inner space. Originally called Micro-World, the attraction would carry guests in Omnimovers (which were then called Atomobiles) into the miniature world of a snowflake.

The storyline would have guests getting smaller and smaller during their journey, until they could actually see the atoms of oxygen and hydrogen that make up water as the snowflake starts to melt. In order not to be lost forever, the guests would be quickly returned to normal size and the real world. During their return, they would see a huge moving eye in a microscope observing their transformation.

The final version of this attraction opened on January 27, 1967, and was called Adventure Thru Inner Space. It was one of the most popular rides at Disneyland until it closed almost 20 years later on September 2, 1985. (The space was needed for a new attraction: Star Tours.)

One of the most memorable films of the 1960s was *Fantastic Voyage* (1966), a combination of the Cold War spy films popular at the time and a clever science-fiction idea about exploring the inside of the human body.

A team of specialists on board a miniaturized submarine were injected into the body of a defecting Russian scientist who suffered a dangerous blood clot during an attack by Russian spies. In addition to dealing with the blood clot, they have to battle the body itself, in particular white blood cells.

Fantastic Voyage was a big hit for 20th Century Fox and was highly praised for its special effects, earning two Oscars: one for Best Art Direction and one for Best Special Visual Effects. (Some critics claimed the best visual effect was a young Raquel Welch in a skintight diving suit.)

It was the inspiration for the 1987 Joe Dante film comedy *Innerspace* with Martin Short and Dennis Quaid where a miniaturized pod is injected into a hapless bystander.

The Disney Imagineers were intrigued by this concept of exploring the inside of the human body in a miniaturized submarine vehicle. However, the crudeness of 1960s technology prevented Adventure Thru Inner Space from being more than a charming attraction where the styrofoam snowflakes often fell victim to guests poking at them, hitting them with a baseball bat, and even shooting a BB gun at them.

That experience taught Imagineers that there needed to be sufficient space between the guests and the attraction scene.

Adventure Thru Inner Space did become a popular "date night" ride. The darkness surrounding the enclosed Omnimovers allowed the opportunity for guests to become more friendly than on other attractions.

When it finally closed, Disney Guest Communications received an irate letter from a couple who complained about the attraction closing: "How could you close that ride?" they wrote. "Our son was conceived on that ride!"

The advanced technology of Star Tours inspired the Imagineers to once again try developing an "inner space" attraction of a submarine-like probe journeying through a patient's body for the Wonders of Life Pavilion at Epcot in 1989.

Body Wars opened 22 months after Star Tours at Disneyland. By 1994, surveys showed it was the most popular ride at Epcot.

The storyline was packed with excitement about guests entering MET (Miniaturized Exploration Technologies) Labs to participate in an experiment. (The sponsor for the pavilion was MET Life Insurance, so it was a clever play on words.)

The probe's captain, Jack Braddock (Tim Matheson from *Animal House*), was setting out on a fairly routine medical mission with a crew of civilian observers accompanying him. The submarine (Bravo 229) and crew were miniaturized to the size of a single cell and beamed inside the human body to rendezvous with Dr. Cynthia Lair (Elizabeth Shue, who starred Touchstone's 1987 film, *Adventures in Babysitting*), an immunologist who also has been miniaturized to study the body's response to a splinter lodged beneath the skin.

Unfortunately, the mission becomes a high-speed race against time when Dr. Lair is swept from the splinter's site into the rush of the bloodstream.

Through the pounding chambers of the patient's heart and through the lungs' gale-force winds, the ship rides the body's current in an effort to rescue Dr. Lair. Even after she's safely on board, there are still problems when the ship loses power and must head toward the brain in search of emergency power to help them escape.

Every guest boarded Bravo 229, but there were other submarines listed to create the illusion of a functioning fleet: Zulu 714, Sierra 657, Foxtrot 817, and Charlie 218.

The film for the attraction was directed by Leonard "Mr. Spock" Nimoy, who had recently finished directing Touchstone's *Three Men and a Baby* (1987). With anatomical images produced by computer graphics and special-effects film techniques, it was a remarkably realistic experience.

Nimoy said:

> Even though Body Wars is the shortest film I've ever directed, it presented a new set of challenges. We had to take into account that the film will be shown inside a moving theater—the simulator. So, in order to intensify the sense of motion, we built a set that actually moves, and rocked it during filming to match the pitching and rolling of the simulator.

Since the story of the attraction was that you were in the bloodstream, the Imagineers programmed movement to mimic the beat of a pulse. That subtle additional movement, which does not happen in Star Tours, may have been what unsettled me as well as others.

Some people contend that it was the gooey images of the inside of a human body that contributed to their uneasiness. It is common for some people to faint at the sight of blood and for even beginning medical students to get ill during a dissection or anatomy class. Some guests had a similar reaction, and the film was eventually shortened from its original length because of numerous complaints.

If the film slipped out of synchronization even slightly, that could also cause a feeling of uneasiness. Although there is debate as to the exact cause or causes of simulator sickness, a primary suspected cause is inconsistent information about body orientation and motion received by the different senses, known as the "cue conflict theory". For example, the visual system may perceive that the body is moving rapidly, while the vestibular system perceives that the body is stationary.

The inside of the ride vehicle, or "probe", was constantly monitored, so if Disney saw a guest experiencing discomfort, they could immediately shut down the attraction. Hopefully, they would shut it down before it got to the point of "Code V", the Disney term for a guest who has vomited.

Unfortunately, on May 16, 1995, a four-year-old girl named Linda Elaine Baker from Texas slumped over in her seat three minutes into the ride. She was seated next to her single mother. The ride was immediately shut down and paramedics were called. Two nurses visiting from Germany were on the ride and tried to revive the girl using CPR. She died after being taken to Orlando Regional Medical Center. The ride was shut down for an investigation. It was determined that the girl had a pre-existing heart ailment and an autopsy revealed that it had not been aggravated or triggered by the movement of the ride.

One Disney secret was that if attendance was slow, they could "lock down" a probe so someone like an expectant mother could sit and enjoy the film without the jarring movement. With a total of four probes available, that was always a possibility.

Like Star Tours, each probe could hold 40 guests. The probes were self-contained units that weighed approximately 20,000 pounds empty and about 27,000 pounds when loaded with guests. They were approximately 10-feet high, 17-feet wide, and 26-feet long. The speed varied depending upon the axis of motion. In its starting position, it was approximately ten feet off the ground.

It was almost like riding in a building. The air conditioning, film, and sound were individually controlled by each flight simulator. The ride itself was cued by the film. Each frame of film generated a time-code pulse with an associated set of jack positions.

When the ride was being developed, an Imagineer watched the film repeatedly while moving a computer joystick to indicate movement and to synchronize the ride and the film. The 70-millimeter film ran at 24 frames per second.

The entrance and exit ramps were "photographed" by an infrared beam after each load/unload cycle. The beam acted as an intrusion system and could sense something as light as a piece of paper on the ramp so that the ride wouldn't move until everything was clear.

The simulator (Rediffusion ATLAS-Advanced Technology Leisure Application Simulator) consisted of a cabin supported by six servo-actuators. In some ways, it resembles the All Terrain Attack Transport ("Walkers") in the *Star Wars* movies.

The actuators were powered hydraulically and driven automatically using electrical drive signals received from a free-standing motion-control cabinet. The actuators provided "six degrees of freedom movement" so the cabin could be moved in planes representing heave, surge, and sway, and in axes representing pitch, roll, and yaw, independently or in any combination.

By June 2001, MetLife had ended its sponsorship of the pavilion, but it remained open until 2004, when it became seasonal. Closing on January

1, 2007, after the holiday season, Wonders of Life never reopened as the health pavilion.

With the 2007 Epcot International Food & Wine Festival, the pavilion became an event facility.

Although there were rumors that the simulators would be shipped to Hong Kong Disneyland for their proposed Star Tours attraction, that never happened.

There are more credible reports that the simulators were stripped for parts to use in the Star Tours attraction at Disney's Hollywood Studios, so it would be difficult for Body Wars to be brought back to life and rush through the circulatory and respiratory systems and make guests queasy.

All Aboard the Fort Wilderness Railroad

The Fort Wilderness Resort and Campground was a unique experience from the day it opened on November 19, 1971, with campsites for guests to stay during their visit to Walt Disney World.

Fort Wilderness (named after the fort on Tom Sawyer Island at Disneyland at the time) was larger than most campgrounds, so trams, bicycles, and buses provided guests with transportation to get where they needed to go.

Guests were forbidden from driving their own cars to locations inside the campground itself because of lack of parking, even if they just needed to go to the Meadow Trading Post to quickly pick up something.

With Disney rushing to open the Magic Kingdom on time, and with fast approaching deadlines and budget overruns causing the elimination of many proposed features, it turned out that Fort Wilderness was just another victim of opening without everything being completely in place.

From the beginning, there were plans for a "campground railroad" to provide transportation and add to the rustic "theming" of the area, but it was not considered a priority, so it did not debut with the opening of the resort.

A clever example of re-using existing assets to save money for the resort was demonstrated by installing trash cans designed to look like tree stumps that had previously been used at the Indian Village area at Disneyland before they were shipped to Fort Wilderness.

The railroad was not solely for transportation, because free trams and buses continued to operate even after it was in operation. The railroad was considered a Disney attraction, and was promoted accordingly on marketing material, even charging guests a minimal fee to use it.

This was the only Disney resort, so far, that had an attraction. It lasted roughly six years and one month, and images of it in operation appeared in magazines and guides, and on posters.

The railroad consisted of four steam trains, each pulling five cars around a circular route through the campground at a maximum speed of

10 miles per hour. Each engine ran on steam and used diesel fuel to stoke the fire. The track was approximately twice the length of the track at the Magic Kingdom.

The strikingly beautiful train was decorated in a strong color palette of forest green, red, and gold. Imagineer Bob McDonnell created this color scheme along with the distinctive logo, detailing, and attraction poster.

A single train was roughly about 150 feet long and could seat up to 90 guests. The trains were smaller than the ones at the Magic Kingdom and were based on the traditional Baldwin "plantation locomotives" popular in the Hawaiian islands, where they were used to haul material like sugar cane and pineapples through the jungles to market.

The railroad used a smaller gauge track (30 inches between the rails on the track) than at the Magic Kingdom (36 inches), which may have influenced people into thinking that the train itself was scaled smaller. The engine was full-sized, not the 4/5th scale usually cited, but it was smaller than the engines operating at the theme park.

Smaller gauge was common for railroads used in logging because it helped the locomotive negotiate tighter curves, just like the narrow turns at Fort Wilderness. So, the smaller gauge and smaller size of the engines gave the illusion of the railroad being smaller than a "real" railroad when, in fact, it was indeed full size for the type of railroad it was.

Unlike every other Disney train (even the ones operating on Big Thunder Mountain Railroad), none of the engines were named. They were only numbered, and each of the four engines had a distinctive icon on the headlamps: elk, bison, deer, and ram.

During 1972, the locomotives were built in Glendale, California, by WED (Imagineering) and its manufacturing arm, MAPO, at a reported cost of more than $1 million. Simultaneously, Buena Vista Construction was installing the track at the campground.

While it was assumed that this was done to save time (like the building of hotel rooms being done separately from the construction of the framework for the Contemporary Resort), later research has confirmed that it was merely a cost-saving effort.

Buena Vista Construction had no experience in laying track, which caused major problems that eventually led to the death of the railroad. Lack of gauge rods and tie plates, incorrect placement of ballast, rails not curved with a rail-bender, and the wrong size rail immediately caused operational problems that continued to plague the running of the railroad throughout its existence.

Imagineer Bob Harpur, the railroad's technical foreman in Florida, told historian Michael Broggie:

The money spent putting the track in was minimal. The roadbed wasn't properly prepared. We had trouble maintaining the track. The locomotives were sensitive. Other than that, they ran fine.

Roundhouse foreman George Britton and his crew tried to fix some of these issues from January 1976 to May 1976 before the opening of River Country, but, despite their efforts, the basic problems still remained, because the money was not available to re-do the entire track correctly, just some "quick fixes".

The trains were shipped by flatbed trucks to Disney World during the spring of 1973 and went through a trial-and-adjustment period, during which guests were sometimes able to ride the trains. By late November, the trains started running on a fairly regular schedule in what was considered a "soft opening".

The official opening and dedication ceremony was on January 1, 1974. Mickey and Goofy, along with the Disney World ambassador, drove a "golden spike" into the rails to declare the train line operational for all guests.

Originally, the train was free for guests staying at Fort Wilderness, with others (including guests staying at other Disney World resorts) charged fifty cents per day (two cents of that was for taxes). The fee increased in later years to a dollar.

In the beginning, the train ran from 8am to 9pm daily, drawing complaints from some guests who disliked the fact that at all grade crossings the extremely loud whistle would sound. Eventually, the trains would cease operation around 5pm, eliminating that problem.

In 1974, Pioneer Hall (with the Hoop-Dee-Doo Musical Revue) and Treasure Island (later to become the bird sanctuary Discovery Island) opened, increasing guest traffic at Fort Wilderness.

However, the heyday of the railroad was the opening of River Country in May 1976, where the train became the favored mode of transportation. The railroad ran three trains at once and, during the busy times in the morning and afternoon, added a fourth train. At one point, the Disney Company tried adding an additional sixth car to each train to increase capacity, but with all trains operating that addition was quickly abandoned as unnecessary.

New cast members were added to help operate the four trains. Six full days of training were devoted to each new cast member, often running into the evening hours. Running a railroad was much more complex than operating the typical theme park rides.

While the first female fireman on the Disneyland railroad did not appear until the 1990s, roughly half of the crew members of the Fort Wilderness Railroad in the 1970s were women. They proved themselves to be more reliable and diligent than some of their male counterparts.

With the opening of River Country, a new addition did have to be made to the train cars: rubber floors, because of the dripping wet guests who had enjoyed Disney's first water park.

Why did the railroad close? Over the years, in lieu of the Disney Company giving an official explanation or even an official closing date (the railroad was just put on "hiatus"), there were many speculations.

Some claimed that safety was an issue and that the nearness of the tracks to the guests made Disney Legal fearful. Some claimed that the train produced too much noise and it disturbed guests. Some claimed that there were challenges with the engines that required constant refilling with water. Some claimed that it was just too expensive to operate and could never recover its costs.

The bottom line is that since the track was not laid correctly in the first place, even with subsequent attempts to make adjustments, the basic problems still existed and could not be overcome.

The entire length of track would have had to be re-laid correctly, at an estimated cost of over $3 million. At one point, the Disney Company looked to General Electric as a possible corporate sponsor to help defray all or most of that cost.

Dick Nunis, who was then n charge of Walt Disney World, loved the railroad and made every attempt to save it by trying to manage costs. In the early 1980s, he brought in the Nature's Wonderland trains from Disneyland that were no longer in use because they had been removed to make way for Big Thunder Mountain Railroad. These were battery powered and so would save fuel costs and had proven dependability. Unfortunately, the train was much too slow on its trial run around the campground and the plan was abandoned.

The Fort Wilderness Railroad almost came to life again in 1992 with the announcement of the building of Buffalo Junction (sometimes referred to as Wilderness Junction).

Before the building of Disney's Boardwalk, Disney had plans to build an upscale 600-room resort between Fort Wilderness and Wilderness Lodge. It would have been themed to the Old West of Dodge City. One of the main attractions would have been a duplicate of Disneyland Paris' Buffalo Bill's Wild West Show. The bottom level of the buildings would have shops and restaurants, while the upper levels would have rooms for guests. All three resorts would have been connected with a new version of the Fort Wilderness Railroad.

The story would begin at Fort Wilderness, representing the original frontier period of the United States. Wilderness Junction would showcase the expansion to the Wild West, and finally, the story would end at Wilderness Lodge, where Americans now live in harmony with their environment. The

Villas at Wilderness Lodge were supposed to represent the housing for the workers who built the lodge and the railroad.

In the September 2010 issue of *The Orlando Business Journal*, it was reported that the same area might be developed into a Disney Vacation Club resort with the possibility that the Fort Wilderness Railroad might be revived in some form. There has been no further news since that speculation.

After years of being outside and subjected to Florida heat and humidity, the engines and the coach cars were sold off to private collectors who restored them. All of the engines are now in California. Disney Executive John Lasseter has an engine and a couple of coach cars in his backyard railroad in northern California.

Two of the coach cars were modified and placed temporarily at the entrance of Pleasure Island as ticket booths. One of those coaches is now at the front of Typhoon Lagoon, and the other was auctioned off. Four of the cars and 3,000 feet of track were donated to the Brevard Zoo in Melbourne, Florida, but over the years those coaches found other homes.

The look of the Fort Wilderness Railroad coach cars inspired the street cars for Disneyland Paris designed by Imagineer Bob Harpur, so the spirit of the original railroad still lingers in a Disney theme park

Ken Anderson's Haunted Mansions

When the topic of the Haunted Mansion comes up, most Disney fans immediately think of Marc Davis. More astute fans may be able to reference the names of Yale Gracey (responsible for the effects) and Rolly Crump. Fewer can recall Claude Coats, the person responsible for the attraction's scary tone.

No one seems to mention Ken Anderson.

While all those cited are deserving of the attention they receive for their contributions to the iconic attraction, Ken Anderson is the "forgotten" man who laid its foundation. His early work on the project strongly influenced some of the things we all love best about this happy haunting ground.

According to Anderson's storyline, a well-known and feared pirate captain quietly retired to private life in a seaside community, like the famed Captain Henry Morgan. He changed his name and used some of his ill-gotten booty to establish himself as a respected and prosperous man. To make his life even more complete, he chose a lucky eighteen-year-old to be his bride and bear him many children.

The only restriction he gave her was to stay out of the attic of their magnificent mansion. Of course, the curious girl couldn't resist and, on their wedding day before the ceremony, but dressed in her wedding gown, she snuck up into the cluttered attic and found a locked trunk that she forced open.

Inside the trunk were souvenirs and documents from the man's previous life. The pirate captain caught her in the attic and, enraged that his secret might be revealed to the community by this foolish girl, he tossed her out the window to her death.

Riders actually relive this experience of being tossed out the window. Next time, look to your right as you leave the attic through the window and you will see that the shingles on the roof of the mansion do not match the ones on the outside of the mansions at Disneyland or Walt Disney World. Like a medium, you have become the girl and see what she saw, including the house from which she fell to her death.

At the bottom of the drop, the scared caretaker of the graveyard is not looking at the ghosts in the cemetery. He is looking at you because you have become a new ghost to join all the others. That is why he is so scared.

The girl's ghost haunted her fiancé so mercilessly that the only way he could find peace was by hanging himself. However, their passions were so intense that their spirits were bound to the mansion for all eternity. Their continuing struggle, even after death, attracted other ghosts, including some who came to celebrate a wedding that will never be held.

Of course, with the redesign of the attic and a change of the storyline to make the bride (now the "Black Widow Bride") a mass murderer of multiple husbands, that tale is no longer intact. Disney theme parks represent a living story that continues to evolve from what the original Imagineers intended.

Anderson (1909–1993) had a bachelor's degree in architecture and intended to pursue a career as an architect, but kept getting sidetracked. He was employed at MGM Studios as a sketch artist working on films like *The Painted Veil* and *What Every Woman Knows*.

One day, as he was driving by the Disney's Hyperion studio in early 1934, he went in and applied on a whim, even though he told his wife, Polly, that he didn't know how to draw cartoons. Impressed with his portfolio, Disney hired him, and Anderson began his career at the studio in 1934, contributing to many animated classics as art director beginning with *Snow White and the Seven Dwarfs*.

Since he had an architectural background, he was able to come up with innovative perspectives on such Silly Symphony cartoons as *Goddess of Spring* and *Three Orphan Kittens* that had never before been used in animation.

Disney historian Paul Anderson (no relation) who has hours of taped interviews with Ken Anderson, once wrote that:

> At Disney, he showed a versatility that has long since gone unmatched. He did so many different things for Disney, that it prompted Walt once to call him "my jack-of-all-trades".

Ken Anderson's official credits list such titles at Disney as art direction, art supervision, story, color, styling, layout, production, and character development. He worked on the classic scene of the dwarfs dancing with *Snow White*, but it became apparent that his talents lay not in animation, but in such areas as production design on such films as *Sleeping Beauty* and *101 Dalmatians*.

Anderson was the one who okayed the use of xerography in the later film which led to a temporary rift with Walt Disney, as well as Anderson's first heart attack when he felt he had failed Walt. Xerography was supposed to save costs of inking the cels and make the artwork on the screen more closely approximate the work of the artists. Walt felt the final result was too sketchy and "arty". He wanted animation to look realistic and smooth.

Specializing in character design in his later years, Anderson designed such characters as Shere Khan in *The Jungle Book* and Elliott in *Pete's Dragon*. He told publisher Steve Fiott:

> I don't know how I came up with Elliott. I like to think of him as an example of China's concept of the dragon as a symbol of luck and good will, which come to them when they need him. He just came to me, and I sure needed him!

It is Anderson's innovative character design on the animated feature *Robin Hood* (which basically was an animal head placed on a human body covered with fur) that inspired an entire generation of young artists, including a group of character costume builders known as "furries".

Anderson also designed many parts of Disneyland. Among his accomplishments: major portions of Fantasyland, like the early dark rides of Snow White's Scary Adventures and Mr. Toad's Wild Ride; the Storybook Land Canal Boat experience; and others.

He retired on March 31, 1978, but continued to consult at WED Enterprises (now Walt Disney Imagineering). He was honored with the Disney Legends award in 1991. One of the last contributions Anderson made to Disney was on *Catfish Bend*, a proposed animated feature for which he made some preliminary sketches. All of that work resides in the Disney vaults waiting for someone to find it again.

Anderson was 84 years old when he died. Forty-four of those years had been spent in the service of the Disney Company.

It had always been part of Walt's vision for Disneyland to include a haunted house, but he had originally planned to locate it on a side street off of Main Street. Imagineer Harper Goff had sketched a dilapidated old house on the top of a small hill overlooking a church and a graveyard. Guests would have looked through its large windows to see the ghostly activity inside.

By 1957, when Walt gave an interview to the BBC about his plans for a retirement home for ghosts, he envisioned it in an area of Frontierland with a New Orleans theme. Ken Anderson had just moved over to WED Imagineering from the studio, and Walt turned to his "jack of all trades" to research some possibilities. As Paul Anderson told me:

> We always hear about "the group of Imagineers" that went out to research the Haunted Mansion. Actually, "the group" was just Ken.

Anderson's 1958 sketch of a decaying mansion was apparently based on the Evergreen House in Baltimore, Maryland. It's the sketch that Sam McKim transformed into the famous concept painting.

It is also apparent that Ken's approach to a Disney "walkthrough" haunted house was greatly influenced by his experiences taking a tour at the Winchester Mystery House in San Jose, California.

Sometimes called the "ghost mansion", this popular tourist attraction has doors and staircases that lead nowhere and a maze of rooms constantly being redone by the widow of the maker of the Winchester rifle. She believed that the ghostly victims of the Winchester rifle had cursed her family and were haunting her to keep building more and more rooms for their earthbound spirits. It was even rumored that through "automatic writing" she received building directions which she passed along to the carpenters.

Ken Anderson had two pages of notes on the Winchester House, from the size of the tour group (maximum of 20), the mix of adults and children (roughly four times the number of adults to children), the maximum/minimum entrance and exit time in each area (25–60 seconds), the maximum/minimum time the guide spoke in each area (32 seconds to three-and-a-half minutes), as well as a variety of notes like "average group well behaved" and "rooms are all empty—nothing to touch".

Anderson strongly believed that a cohesive story was necessary to guide Disney guests through the Haunted Mansion. Not only was storytelling an important element of the Disney Brand and had positioned Disney's "dark rides" as different from carnival amusement park dark rides, but storytelling would be necessary to move guests through the experience rather than have them dawdle in one area and clog the flow of traffic.

In 1957, Anderson developed four story concepts, all of which feature elements reflected in the final version of the attraction that was eventually opened at Disneyland on August 9, 1969.

Perhaps the best-known version was Anderson's first attempt, which featured the Legend of Captain Gore and was written in February 1957. The mansion was the seaside manor of an old sea captain who had married a young woman named Priscilla. She discovers he is actually a notorious pirate.

Gore killed his bride (and in one version tossed her bloody body into an outside well that still bubbles red) and she haunted him until he took his own life by hanging himself. This story would have been shared with guests by a butler or maid who worked in the mansion.

Another version was Bloodmere Manor, which was the lakeside estate of "the unfortunate Blood family". It was built around 1800 in the swampy bayous near New Orleans and was moved to Disneyland intact because it was an example of early architecture from that region.

The mansion had not been occupied for some time and was badly in need of repair, so the Disney Company started the work of restoration as soon as it arrived at Disneyland, but:

[S]trangely enough...the work of each day was destroyed during the night...and the night watchman reported that when he had passed the house he'd heard eerie screams and seen weird lights... In fact, we are sorry to report that the latest tragedy of all occurred here in Disneyland...when one of our carpenters engaged in restoration work on the house disappeared completely from sight...and he has not been seen or heard from since. The house is now too dangerous to live in, but we have succeeded in making it safe enough for a visit... when accompanied by our trained and competent guide, a former butler of the household.

A third version had Walt himself as the narrator on tape as the guests wandered through the house to go to a ghostly wedding celebration.

Another version focused on the Headless Horseman from the Disney animated film *The Legend of Sleepy Hollow* (1949) about Ichabod Crane on Halloween night. This version also featured a wedding, this time between Monsieur Bogeyman and Mlle. Vampire. The bride jilts the groom at the altar, sparking chaos and the need to quickly exit the mansion.

Let's take a closer look at Ken Anderson's second revision of the Bloodmere Manor version of the Haunted Mansion, dated September 17, 1957. It is one of the most detailed (24 pages, double-spaced) and atmospheric of the versions that Anderson submitted. Here is the first page:

Guests will be admitted to the grounds through a large wrought-iron pedestrian-and-vehicle gate typical of New Orleans, circa 1800. The ticket booth will be located in the brick and plaster gatehouse which terminates the wrought-iron fence. Posted conspicuously on the gatehouse are copies of the *Times Picayune* and *Leslie's* magazine, with headlines about atrocities connected with the ghost house in the past.

The approach to the house will be along paths lined with azaleas and moss-festooned magnolias and southern oak trees. The garden shows evidence of a once well-planned symmetry and beauty, but is now overgrown and obviously out of control.

Vines and moss combined in the tall trees shut out much of the sunlight and lend mystery to the shadowy exterior of the house. Being set well back from the street behind the grove of trees, the house will be scarcely visible until close upon it. It appears to be in a state of dilapidation common to all ghost houses. First at one upstairs window and then another, a girl's face appears momentarily, screams, and is throttled by a large hairy hand which draws her back into the darkness.

Notations on this revision also indicate that Anderson had used his experience at the Winchester House to plan audience flow. He estimated that a group of no more than 40 guests would gather on the front porch to enter the house. At regularly spaced intervals of one-and-a-half minutes, around 8 groups of 40 guests each (320 guests total) could be in the house simultaneously, or a possible 16,000 visitors in a 10-hour schedule.

Anderson wrote:

> If the show in each room lasts a minute, it would leave 15 seconds to enter the room and 15 seconds to clear the room. We are conducting tests with groups of 40 people, using the *Zorro* sets [on the backlot of the Disney Studio in Burbank], to determine the practicality of this timing.
>
> In room clearance tests so far, times ranged from 15 seconds to 25 seconds, for an average of 20 seconds. As soon as construction of the test mock-ups for optical illusions are completed, we will utilize them for further crowd capacity tests; which will include a one-minute show of the illusions. A tour of the house should take an average of about 12 minutes to complete.

The first episode of Disney's popular weekly television program *Zorro* appeared on ABC in October 1957, but workers had started building the sets in June 1955 and they were the Disney Studio's first permanent sets, costing more than $100,000.

(You may recall the so-called Disney urban legend that one of the things causing the delays of the opening of the Haunted Mansion at Disneyland was that it was too scary for test audiences. Other scholars aptly pointing out that the mansion was just a hollow shell, so test audiences couldn't have experienced it. Apparently, there *were* test audiences going through the experience as early as 1957 on the Disney backlot. I wonder if something unfortunate happened which resulted in the birth of that urban legend?)

For the front porch of the mansion, Anderson had written a one-and-a-half minute speech to be recorded by Walt Disney, which would have explained the strange history of the house and ended with Walt explaining:

> The guide is made necessary by another strange characteristic of the house. It has rotted so long in the dank fastness of its lost hiding place in the swamps that not even southern California sunshine or the best efforts of electricians and illumination engineers can dispel the dimness of the bayous...it mysteriously remains always night within the house...the night in which all ghosts are condemned to live.
>
> Now we cannot promise you that anything at all will occur during your visit...since it is usually in the wee small hours that the departed ones live it up... However, be prepared to see and hear something or other, and take whatever precautions you please. We recommend that you stay close together during your visit and please...above all... obey the guide's instructions. Thank you.

Using Anderson's own notes, as well as excerpts from interviews I did with him, we will go on a step-by-step walking tour of Bloodmere Manor, one of the earliest prototypes for the final Haunted Mansion. This information is from Anderson's September 17, 1957, final draft.

In the proposal, the Disney Company has relocated an authentic 1800s mansion from the swampy bayous around New Orleans to Disneyland, and strange occurrences have happened, including noises and people disappearing, but it will be safe enough as long as guests stay close to their butler guide.

Once inside the house, the guests would immediately experience some sound effects and even an invisible ghost writing on the wall: "Foolish mortals... go home!" However, a butler appears and assures the guests they don't need to worry, "He's only a ghost writer!"

This is the first use of the term "foolish mortals" in connection with the mansion.

The butler explains that the last group he took through had the good fortune to see a good deal of ghostly activity. As the butler talks, a panel in the wall behind him opens and a "huge, hairy arm gropes menacingly", but the butler easily avoids it and warns the guests not to get too close to the walls and to stay with the group at all times during their visit "because Hairy the Arm, who was the insane brute of a man-servant for the old Blood family, delights in picking off stray visitors".

The group then passes through a gallery of portraits leading into the Library. The butler instructs the guests to come to the dead center so he can describe the portraits of the infamous ghosts because:

> [Th]e unfortunate Blood Family, which inhabited the house in life, had a tremendous circle of acquaintances and an international reputation as hosts.
>
> The supreme tragedy of the house occurred while the Blood Family was hosting numerous friends on the eve of the real-life wedding of their beautiful daughter. An event too horrible to mention prevented the wedding and it has been rumored that, on every anniversary for the last 150 years, the ghosts have been attempting to complete the ceremony which would lift the curse on the house.

An amazing thing about these portraits is that at first glance they appear quite normal, "but on second glance, the eyes of each portrait appear to stare back directly at the viewer and follow him relentlessly wherever he moves". (Does this concept sound familiar?)

The guests are then led into a huge, dimly lit library. As the butler points out some of the items in the room, the group is joined by the "Lonesome Ghost" who is shunned by other ghosts because he likes people better than his ghostly peers. At first, the group only hears the ghost, but then the butler directs them to look at the huge mirror hanging over the fireplace and the guests see not only their own reflection, but the image of the

Lonesome Ghost apparently moving through the group as he speaks. (A version of the Pepper's Ghost illusion.)

The Lonesome Ghost is excited because "two of our ghosts from prominent old ghost families are getting married" today. The ghost directs the guests' attention to more portraits in the Library. As he talks about each portrait, an eerie light illuminates it and the portrait changes.

Here are some of Ken Anderson's suggestions from 1957:

> As a typical example of the type of reaction, a portrait of the blue-blooded relative will seem to fill up with blue blood like a bottle filling with liquid, sound effects and all. Also, a maiden aunt with an austere face will coyly wink and the portrait of a gay blade will disintegrate ala Dorian Gray, etc.

While my admiration for Marc Davis is boundless, Ken Anderson's descriptions of humorously changing portraits was suggested before Davis even became involved with the project.

Hairy the Arm makes another grab for the butler, who tells the group they will need to wait in the room for a bit, but the Lonesome Ghost suggests another alternative and suddenly the group hears his voice from an adjoining room beyond the walls. "Oh dear me," moans Lonesome, "I forgot you mortals can't walk through the walls... You'll have to use the bookcase..."

A bookcase creaks open allowing the guests to enter a room called the Gallery where they get to experience a screaming female ghost whose head separates from her body. It is the ghost of Anne Boleyn who was beheaded by her husband. "Of course, she'll have to pull herself together in time for the wedding," remarks Lonesome who will now disappear from the group for a while when the guests start to leave.

As the butler tries to lead the group out a large double door, the shadowy figure of Hairy the Arm is seen as it throws a knife which sticks in the wall across the room, forcing the guide to take the group through a secret wall panel.

In this new room, the butler is on the bedroom level when the group arrives at one end on a raised platform separated by a railing from the main floor. Originally, this platform was only a foot or two above the rest of the room, but now it has three separate levels so all the guests can see equally well into the bedroom.

There are a series of ghostly gags that take place here, including "five hideous little Charles Addams-type children monsters who sit up in bed and glower at the visitors as they chorus: 'EEK! PEOPLE!'. A door opens revealing an invisible bather taking a bath using a visible sponge, brush, and wash cloth, and singing 'I'll Be Glad When You're Dead You Rascal You'..." (It's our old friend, Lonesome, getting prepared for the wedding.)

During these displays, Hairy the Arm pulls the butler into the wall and the room goes pitch black with only the sounds of fighting and then silence. The lights come back on with a disheveled butler appearing to usher the guests into a large oval room called the Salon.

The room slants toward a large bay window that displays a windy moonlit scene of distant bayous. As the guests watch, clouds obscure the moon, there are flashes of thunder and lightning, and ghostly skeletons seem to rise from marble tombs and float toward them. There is a distant sound of pounding hoofs that signal the approach of the Headless Horseman, who is eventually seen galloping through the tops of the small trees and overgrown shrubs.

A fireplace near the guests mysteriously rises and Lonesome reappears to invite them to follow him to the wedding. The guests enter an octagonal-shaped room with rough unfinished walls and ceiling and windows on three sides with broken panes.

Suddenly, the storm breaks outside in full intensity with rain drenching the windows and more arriving ghosts (some with skeleton umbrellas). A series of brilliant lightning flashes reveal the transparency of the ceiling. As it beomes transparent, guests can see at the highest point in the peaked ceiling the ghost of a figure in full dress clothes...hanging by the neck.

Since the room seems to be filling up, the guests are ushered into another short mirrored hallway and as they look in the mirrors they see not only their own reflections, but the transparent reflections of ghostly visitors heading in the same direction.

Anderson wanted some of those ghosts to be famous, like Dracula, Frankenstein, the Hunchback of Notre Dame, the Phantom of the Opera, Scrooge and Marley, Little Eva and Simon Legree, Jack the Ripper, the Canterville Ghost, Captain Hook, King Tut, and so forth. According to his notes, blacklight would be used in this effect, as well.

Eventually, the guests would arrive in a large octagonal-shaped room with mirrors on all sides. Even the doors are mirrored, so that guests can clearly see all the ghosts in the mirrors who are attending the wedding and chattering away about the wedding gifts of a matched set of poisoned darts, guillotine book ends, etc. The Lonesome Ghost mingles through the group of ghosts.

The sound of wedding bells causes all the ghosts to disappear, leaving only the reflection of the guests who move into the great hall, where the wedding is to take place. On the lower floor is a long table with a wedding cake, candles, and flowers, all covered in cobwebs. An invisible ghost plays an old pump organ.

The groom appears, as does the bride on the opposite side who floats toward him. He tenderly reaches out and takes her head off and kisses it.

She retaliates with a resounding slap. This action is repeated several times. This causes lights to flash, thunder claps, rattling shutters. The groom kisses her again and is slapped again.

Suddenly, the music gets faster and faster to crazy rhythm. Footprints run all around the floor and walls below, while furniture is upset and the organ is now joined by various floating jazz instruments. The storm builds to a climax. The ceiling collapses and rain pours in.

As things intensify, the butler leads the guest into the Trophy Room, where the skulls of ghostly animals on the wall stare back at the group as the portraits did in the Portrait Hall. The group is rushed through a fireplace, still hearing the pounding rain and the loud chaos behind them. As they are led outside, they are surprised to see that it is not raining at all. (This is many years before the Enchanted Tiki Room, which uses a similar effect at the end of the show.)

The butler attempts to discuss the Blood Family crypt and graves which the guests still have to visit, but Hairy the Arm grabs the butler and pulls him back into the house with a bloodcurdling yell. The tour is finished and apparently so is the poor tour guide.

Anderson even thought that an attraction should end in an area where guests could purchase merchandise related to their experiences. This concept later became commonplace at Disney theme parks, but here is the first time it is introduced. In the garden crypt:

> [V]isitors will be given the opportunity to buy pieces of ghost wedding cake neatly wrapped in shroud material, and tied with a bow of ribbon...suitable for placing under the pillow for inducing dreams; Lonesome Ghost lapel buttons, which plug a visit to the Ghost House; or Lonesome Ghost balloons, complete with floating China silk shrouds.
>
> In the high-walled and fenced garden, the visitors may take many paths all leading to the exit. There will be knee bones, foot bones, and skulls protruding occasionally from the silent paths. A statue in a secluded spot in the unkempt, overgrown garden animates mysteriously at timed intervals. In one corner of the garden is a typical graveyard with epitaphs to be inspected, while closer to the exit gate is a wishing well with an echo effect.

Anderson moved back from WED to the studio in the late 1950s for a variety of reasons (including that, with the mounting costs on *Sleeping Beauty*, the animation department needed to be brought back under control, which is why Anderson proposed using xerography to save costs on hand inking cels) and spent much of the next several years working on the Disney animated feature films.

In 1964, Walt assigned the task of the Haunted Mansion to Marc Davis, Claude Coats, and X. Atencio, while Crump and Gracey developed a "Museum of the Weird" which would have been a "spill area" near the mansion where guests could enter and exit at their leisure before going through the main attraction, or after they had experienced it.

In later years, Davis was quite vocal about how Walt didn't want a storyline for the Haunted Mansion and that, when Walt died, there was a struggle over creative direction. Most Disney fans agree that the first part of the attraction, with its spooky elements, reflect the contributions of Coats, Crump, Gracey, and even Anderson, while the second part of the attraction, especially the graveyard scene, is indicative of the more humorous approach of Davis, plus some funny elements suggested by Anderson.

This disjointed approach resulted in cast members and guests creating their own cohesive storyline for the attraction using bits and pieces of the various storylines from the original proposals.

Looking over Anderson's concepts from the late 1950s, it becomes apparent that he should receive greater recognition for his contribution to the attraction. Just the short summary of one of his proposals reveals many elements, from the transparent ceiling revealing a hanged gentleman, to portraits which transformed humorously before your eyes, to looking in a glass reflection to see both a guest and a ghost, which are some of the most memorable experiences in the current Haunted Mansion.

All of these ideas were in place before other talented Imagineers were even assigned to the project, nearly five years later. Of course, it is not unusual (or unethical) for Disney Imagineers and animators to be inspired by or to build upon the work of earlier artists.

Who Disappeared Roger Rabbit from the Disney Theme Parks?

Who Framed Roger Rabbit has often been credited as the spark for the birth of an animation renaissance in the 1990s, although today the Disney Company claims that the revived interest was started by a mermaid the following year.

The peak of Roger Rabbit's Disney career was roughly 1988–1992, after which he all but disappeared from the films and the parks. Here is what happened to a character that many thought would be entertaining audiences for many years.

Disney President Ron Miller, against the advice of CEO Card Walker, purchased the rights to the 1981 book *Who Censored Roger Rabbit* by Gary Wolf for $25,000, even before it was published and put it into development.

This was a time of turmoil at the Disney Company and, when the dust finally settled, Michael Eisner and Frank Wells were in charge in 1984, and many previous projects approved during the Miller era were abandoned. However, Jeffrey Katzenberg discovered the work that had been done on Roger Rabbit and shared it with Eisner, who put the project back into development.

When Eisner was head of production at Paramount, he had given Steven Spielberg support to direct a film, after Spielberg's previous cinematic flop had almost made him unemployable in Hollywood. That Paramount film turned out to be *Raiders of the Lost Ark* (1981) and generated a great amount of goodwill between Eisner and Spielberg, whose career was resurrected by its success.

Eisner knew that a Roger Rabbit film would be expensive and difficult, and so to minimize the financial impact to the Disney Company, he made arrangements for Spielberg to help with the production.

Spielberg negotiated a contract that resulted in Amblin Entertainment (which he had co-founded in 1981) getting not only major creative control, but also 50% of box-office receipts, licensing, merchandise, theme park attractions, and just about everything else. Any project involving the original characters, including Roger Rabbit, Jessica, and Baby Herman, would require mutual approval from both Amblin and Disney, which had joint copyright of the film.

Eisner anticipated no problems. He and Spielberg had a friendly relationship, and while Spielberg had an interest in animation (just having completed the successful Don Bluth production *An American Tail* in 1986), Eisner figured he would never be so deeply involved as to become an animation or theme park competitor and, most importantly, this was an opportunity to establish a business relationship so that Amblin might supply future live-action films for Disney.

After the film was released, Eisner felt that Spielberg would "rubber stamp" anything Disney wanted to do with Roger and friends, because Spielberg would be more concerned with live-action productions.

Eisner's false assumptions would result in the disappearance of Roger Rabbit at the peak of his popularity.

Spielberg's contributions to the success of the film were significant, including convincing Robert Zemeckis (hot off the success of *Back to the Future* in 1985) to direct, negotiating with individual cartoon studios to loan out their famous characters for an appearance for the paltry price of $5,000 per character, arranging for Industrial Light and Magic (ILM) to do special effects, and setting up an independent temporary animation studio in the United Kingdom under the supervision of Richard Williams (because the feeling was that Disney could not handle the demands of the necessary animation).

Eisner felt he had wisely hedged his bets and, according to his autobiography, "we had hoped to distance it [the film] from the Disney brand by releasing it under the Touchstone label" because there was the real fear that it might turn out to be another *Howard the Duck* (1986) disaster (Howard was also a comic book character in a human environment) or even a public relations nightmare because of some of the controversial elements in the film that Eisner considered too "sophisticated and sexy".

All of those fears were alleviated when the film premiered to critical (especially for the technological innovations in the pre-CGI era) and financial success (it was the second-highest grossing film of the year behind *Rain Man* and brought in more than $300 million on its $50 million investment), earning three Oscars and a special Academy Award for Richard Williams, as well as recognition at many other award ceremonies.

Eisner wrote:

> By the time it [the film] premiered, we had licensing agreements for over 500 products, ranging from Jessica Rabbit jewelry to Roger

Rabbit talking dolls to computer games. Both McDonald's and Coca-Cola created massive promotional tie-ins.

The success of the film inspired Eisner to immediately increase the promotion of the character whom the public perceived as a Disney character, not connected with Amblin.

Who Framed Roger Rabbit was released in the summer of 1988 and, by that fall, a costumed Roger Rabbit character was appearing at Disneyland. In addition, Roger had been included as a major character in an NBC television special saluting Mickey Mouse's 60th birthday, a theatrical short called *Tummy Trouble* (the first new Disney theatrical short in more than 30 years) had been put into production, a possible film sequel was being discussed, and Roger was slated to become an important presence at the Disney-MGM Studios that celebrated the Hollywood of the 1940s since the original film was set in 1947.

Disney fans often forget how prominent Roger Rabbit was at the Disney parks for several years.

Roger Rabbit was one of the six gigantic 45-foot-tall inflatable balloon figures in Disneyland's Party Gras parade from January to November 1990 to celebrate Disneyland's 35th birthday. Walt Disney World brought the parade to the Magic Kingdom for that park's 20th Anniversary Surprise Celebration in 1991, and Roger was given a new jester's hat. The parade ran for approximately three years.

He was featured in several scenes of the *Disneyland Fun Singalong Songs* video, as well as on Disneyland's 35th anniversary special on NBC.

Roger was also the original conductor in the SpectroMagic night-time parade that debuted at Disney World in 1991 (he was later replaced by the Genie from *Aladdin*). In fact, Roger was the official toon representative of Disney World's birthday celebration. He toured the United States, accompanied by perky male and female Disney performers dressed in Roger outfits to publicize the event.

Roger was spotlighted in the Kids of the Kingdom show in front of Cinderella Castle and was also prominently featured in the Mickey Starland stage show.

At this time, the Disney Company realized that Disney-MGM Studios was so hugely popular that it needed to be physically expanded. As part of the plans for the Disney Decade, one idea discussed for Disneyland was to convert the area behind Main Street U.S.A. into a Hollywood Land with a section devoted to Roger Rabbit and featuring some attractions. In May 1991, Disney officially cancelled the project, claiming in a statement to the *Los Angeles Times* that the "proposed construction would come at the same time as development of the proposed Westcot theme park nearby".

These same Roger Rabbit elements were suggested for a concurrent expansion of the Disney-MGM Studios that was to be located approximately where Sunset Boulevard is today. Known as Roger Rabbit's Hollywood (and sometimes as Maroon Studios, the fictitious animation studio where Roger works), it would have been an entire street that looked as if it belonged in Toontown with its wacky architecture. (Later versions of the concept had it located at the end of Sunset Boulevard near the present-day Rock 'n' Roller Coaster.)

An often quoted *New York Times* newspaper article described it as "...a kind of Toontown, where—as in the movie—only cartoon characters may live". The street would be littered with all sorts of surprises like boxes of cartoony TNT, a grand piano dangling precariously over the street, and Roger-shaped holes in the walls. Red Cars would take guests up and down the street, stopping at the Terminal Bar from the movie that would serve as the restaurant for the new area.

The Toontown Trolley attraction would have been a motion-control simulator like Star Tours, though with some differences, including not only a screen in front but on each side. There would be in-cabin effects like Roger-shaped dents when the character crashed into the roof.

Baby Herman's Runaway Baby Buggy would have been a traditional Fantasyland-like dark ride. Based on the incidents in the first Roger Rabbit short, *Tummy Trouble*, where Roger and Baby Herman have a series of misadventures in a hospital, it would have had guests board oversized baby-buggy ride vehicles and careen down stairs, through hospital rooms, around beds and patients, and more. When the attraction was described in the newspapers, some readers angrily complained that there was nothing funny about a hospital and that the Disney Company was being insensitive to both patients (especially scared children) and doctors, especially since the short featured frightening sharp objects and scary mechanical devices.

Benny the Cab was an attraction planned for the area that did get tweaked and built at Disneyland as Roger Rabbit's Car Toon Spin. Benny was transformed into his "twin cousin" Lenny. Imagineers tried to explain the missing Benny by saying that Roger was out driving him at the time, so he was unavailable. Using Benny would have required an additional payment to Amblin.

On opening day at the Disney-MGM Studios, a costumed Roger Rabbit put his hands and footprints into a square of concrete that was placed almost directly in front of the entrance to the Great Movie Ride. It reads: "PL-L-L-LEESE. Roger Rabbit. May 1, 1989."

At one time, the Dip Machine model and the bullet case used by Eddie Valiant were on display in the queue for that attraction.

On opening day, guests could see a huge Maroon Studios billboard in the Echo Lake area. Today, it is faded, but originally it had a bright maroon color and the faces of Roger, Jessica, and Baby Herman were striking, just like the title card at the beginning of a cartoon.

Over the Hollywood and Vine Restaurant, according to the backstory, the rooms were rented to individuals or businesses. One window features a small sign stating "No Actors" because at the time, actors were considered immoral, likely to have wild parties and probably skip out on paying the rent.

A little farther down is a window that originally stated (recently, some of the lettering has chipped off) "Eddie Valiant. Private Investigations. All Crime. Surveillance. Missing Person." There is also a symbol of a magnifying glass with an eye in it, a reference to "private eyes".

Valiant, of course, was the detective in the film played by Bob Hoskins when the producers could not find comedian Bill Murray to offer him the role.

Next to Valiant's window is another window with the silhouette outline of Roger Rabbit bursting through the blinds and the window, just like in a famous scene in the original movie.

Originally, at the lower-left corner, was a sign that read "No Toons" to parody the previous "No Actors" sign, as well as an insider joke since, during this time period, Valiant disliked toons because one had killed his brother. That "No Toons" sign was removed by a Disney executive who didn't understand the joke and insisted that Disney loved toons and that the location should be tested as a character breakfast area.

However, the sign by itself is not the real punch line to the joke. If a guest can extrapolate the angle and the direction of Roger's outline, it goes in a direct line to the backstage building that once housed Disney Feature Animation Florida, which made two Roger Rabbit theatrical shorts, *Roller Coaster Rabbit* and *Trail Mix-Up,* and were preparing for more when trouble arose.

On the side of the Disney Feature Animation Florida building was a painted black silhouette of Roger, the same size and in the same pose as the one from the window.

Bridgitte Hartley, who worked as an animator on *Who Framed Roger Rabbit* and *Roller Coaster Rabbit* at the Disney-MGM Studios, died from cancer in the early 1990s. Around the corner from the silhouette of Roger on the Feature Animation building was a small garden in her memory called Bridgitte's Garden, created by Max Howard, who was in charge of the Animation Department.

In the outdoor paint area of the Backlot Express restaurant was the actual Toon Patrol black paddy wagon vehicle that the weasels drove to do the evil bidding of Judge Doom. It is a real 1937 Dodge Humpback panel truck. Today, the round official City of Los Angeles Toon Patrol decals on the front doors identifying it have been removed. (Inside the restaurant in the indoor office of the Paint Supervisor, on the bulletin board, are some color photos of the vehicle with the decals and some standee weasels.)

Shoehorned into a little corner of the Stunt Men's area of the Backlot Express restaurant was the original working skeletal frame of Benny the Cab from the movie. In the film, Bob Hoskins sat in the driver's seat, holding a rubber steering wheel.

Behind him, and lower, was stunt driver Charlie Croughwell completely covered in a black jumpsuit and wearing a black hood (thin enough to see through) who actually drove the vehicle. When the live-action filming was complete, a colored cel of Benny would be placed over the vehicle obscuring any live-action reference. On the nearby walls are faded photos of how this magic was accomplished.

At the end of the Backlot Tram Tour, where guests could view Jessica's blue car as well as a Red Car from the film in the Boneyard, a pathway of Roger Rabbit's large footprints on the ground led guests into the Acme Warehouse, more specifically the section known as the Loony Bin, or the Acme Gagworks.

There were some interactive props (later incorporated into Mickey's Toontown at Disneyland) like boxes that would make noises when you tried to open them. Things hung from the ceiling such as a net filled with Acme's Ton of Bricks (that still hangs there today). There was also a standee of Jessica Rabbit for photo opportunities. Behind her, the wall had an open-air silhouette outline of Roger Rabbit who had apparently run through the wall. Across from this location was a photo shop where guests could don Eddie Valiant's trenchcoat and be superimposed into a scene with a more realistic 3D version of Jessica or Benny the Cab.

Judge Doom's Dip Mobile from the movie had also crashed through the wall and guests could take photos posing beneath its huge front roller. The Dip Mobile moved around to different locations over the years, including the front of the Backlot Tram Tour (where a red tram now resides) as well as in the Boneyard on the tour, but is now gone from the park.

A plethora of merchandise, especially themed to Jessica, was available for sale in the area. That same Jessica merchandise, along with so much more, was also available at a Jessica Rabbit-themed store on Pleasure Island from 1990–1992. Most memorable was the exterior with a giant 32-foot-high two-sided neon sign of a full figure of Jessica in her sequined gown sitting down with legs crossed and lazily swinging her left foot back and forth.

Of course, because of the contract with Amblin, a new licensing operation had to be set up with all items copyrighted Touchstone/ Amblin Entertainment and the accounting done separately from other character merchandise.

The theatrical short *Tummy Trouble* had been rushed into production and was released with the Disney live-action film *Honey I Shrunk the Kids* in June 1989. It was credited in the industry as having significantly boosted the revenue for the film. Almost immediately, a second short, *Roller Coaster Rabbit*, was made. Spielberg assumed that this short would be attached to his Amblin film *Arachnophobia* (1990), especially since it was the first feature to be released by Disney's new Hollywood Studios.

However, Disney had invested heavily in *Dick Tracy* (1990) and felt the film needed additional help to recover its costs. Eisner insisted that, since Disney was releasing both films, the short should be connected with *Dick Tracy*.

Spielberg fumed quietly, grumbling that as the co-owner of the characters he should have a say in how they were being used and that a Roger Rabbit short should accompany one of his films. Work began on a third short, *Hare in My Soup*, to be released with *The Rocketeer* (1991). The premise was that Roger, Baby Herman, and Baby Herman's mother went to a restaurant. While the mother left the table to powder her nose, Baby Herman followed the chef into the kitchen and violent chaos ensued.

Pre-production work was finished on the short when Spielberg announced he could not approve the cartoon and he had concerns about the script for the feature sequel as well, including that the villain was a Nazi, and so he wouldn't approve it, either.

Eisner realized that there would be even more trouble getting Spielberg's agreement on any future co-productions.

Artist Peter Emsile remembered that he had just completed the image of Roger Rabbit for Disney World's 20th anniversary press kit cover when word came from Eisner to stop any projects that featured Roger. So, 1992 was the end of Roger Rabbit's career.

Disney tried to create its own version of Roger Rabbit named Bonkers D. Bobcat that it would own completely. First appearing in the Disney series *Disney's Raw Toonage* (1992) and then later in his own television series from 1993–1995, Bonkers was an animated bobcat who became a special police officer teamed with a human partner.

The premise was similar to Roger in having a toon in a real world, but it was undercut because the humans and the city were also animated, although in a less extreme way than Bonkers and his friends.

A costumed Bonkers characters made appearances at Disneyland and Disney World. He also performed in the show at Mickey's Starland in 1993—1994. However, audiences did not embrace the new character, despite his physical resemblance to Roger.

In an attempt to patch things up, Disney shifted focus to a different Roger theatrical short titled *Trail Mix Up* that did get Spielberg's approval and was released with the Disney/Amblin co-production *A Far Off Place* (1993), but it did not receive the same attention as the previous shorts.

It was clear that Roger was pretty much dead in animation and in the Disney parks by mid-1993.

The Disney Company philosophy became: Why fight with Spielberg and share the revenues when there were dozens of new Disney animated characters, from a mermaid to a beast to an upcoming lion?

In recent years, there has been some discussion about whether the chilly atmosphere around Roger Rabbit has begun to thaw.

A towering Roger Rabbit stands precariously on top of a barrel of turpentine that could instantly dissolve him in the 1980s section of Disney's Pop Century Resort, which opened in 2003 and which includes other 1980s references, such as Rubik's Cube and the Walkman.

A costumed Roger Rabbit made a brief return (March 25 through Easter Sunday, March 31, 2013) on Main Street, U.S.A. at Disneyland as part of the pre-parade festivities and to lead the guests in the Bunny Hop dance. He had previously made a one-night only appearance during the 20th anniversary performance of Fantasmic! in May 2012, waving from the *Mark Twain*.

Unfortunately, for those wanting the return of Roger Rabbit, it seems for the newest generation of guests who never saw the cartoons or the character in the parks, the question is often not "Where is Roger Rabbit?" but "Who Is Roger Rabbit?"

PART FOUR
Other Disney Stories

Even before the acquisition of franchises like *Star Wars*, Marvel, and the Muppets, the Disney Company had a rich and diverse history beyond films and theme parks.

Many articles could be written on Disney's involvement in everything from comic books to postage stamps to community events and more. Additional articles could be written about the countless colorful individuals who were connected with Disney projects.

This section of *The Vault of Walt* has always been devoted to those off-the-main-road stories where a traveler might decide to pull over and see the largest ball of string in the world or the little museum honoring school lunchboxes. These are the fascinating stories that would not always fit comfortably in the other sections of the book, but help illuminate the bigger picture of Disney history.

Back in 2005, Disney Archivist Dave Smith solved a mystery for me about why Disney in the early 1950s owned the rights to many non-Disney songs. Smith wrote:

> Fred Raphael, who started Disney Records, was trying to build up the Walt Disney Music Company song catalog quickly, and thought that we should be a hit parade competitor, so in the early 50s he added to our song catalog not only "Shrimp Boats", but the popular "Mule Train", among many others. These songs had absolutely no connection to Disney in any way, other than we owned the licensing rights.

I started wondering, what with Walt Disney's great affection for America, why there was never a Disney song about the Fourth of July. There were Disney songs about other holidays, including Halloween and Christmas. It turns out that in 1975 the Sherman Brothers, in advance of the national bicentennial, wrote just such a song, "The Glorious Fourth" (the sheet music for which I own), even though no one remembers it:

There's excitement in the air; It's a feeling we all share,
The day that Yankee Doodle throws his hat into the air!
Nobody's workin' today, you bet,

Family's are out promenadin'
Picnic tables are being set

The day's made for Disney paradin'
Bunting and banners are everywhere
Hot dogs and fresh apple pie

It's our historical proudly uproarical fourth, the fourth of July.

A million sky rockets and roman candles zing zoomin' on high,
Red, white, and blue sparklers and spinnin' pinwheels raise cain in
the sky,
And white flags wave?and the bands all play

You can't be sad if you try? on the bang-up uproarious, flag waving, glorious fourth, the fourth of July

Even the Sherman Brothers can't always compose huge hits, can they?
However, the reason I treasure this copy of sheet music is that there is
a color picture of Walt and this quote credited to him from July 4, 1964,
that I have never seen reprinted anywhere else:

The Spirit of America is never more clearly seen than in those precious
moments of public displays of patriotic feelings. As a child, I remember the intense wonder and awe with which I was left after singing
the National Anthem or after fireworks on the Fourth of July. It is
my hope that these feelings spring eternal in the minds and hearts
of all Americans.

These are the type of oddball items that have been lost for decades in
out-of-sight and inaccessible nooks and crannies. In this section, I try to
preserve some of that invaluable information.

Fred MacMurray:
The First Disney Legend

"I will say the seven pictures I made at the Disney Studio were the pleasantest times I've had in the picture business and I've been around quite awhile," said Fred MacMurray in the 1973 ABC Wide World of Entertainment special, *Walt Disney: A Golden Anniversary Salute*.

MacMurray built a successful career as the ultimate "Mr. Nice Guy" in a series of films in the 1930s and 1940s. Yet, he was also riveting and critically lauded for his bad guy performances in films. He often described himself as a personality rather than an actor, but he played roles ranging from screwball comedy to romance to film noir to musicals in over one hundred films during his lengthy career.

MacMurray was so popular and heroic in an average man kind of way that artist C.C. Beck claimed he used the actor as the model for his comic book superhero, Captain Marvel. Beck had been doing work on a magazine about movie stars. When he had to come up with a design for the character, he used MacMurray's black, wavy hair, bone structure, and cleft chin.

By the 1950s, MacMurray's box office appeal had lessened and he appeared in many low-budget Westerns. His career revived with his appearances in several popular Disney live-action films beginning in 1959, as well as in a successful television series (*My Three Sons*) that ended in 1972.

He disappeared from films for a few years, but resurfaced in another Disney comedy, *Charley and the Angel*, in 1973. After a pair of made for television movies, MacMurray made one last feature, *The Swarm*, in 1978, before retiring.

However, about a decade later, he was pulled out of retirement briefly to become the very first Disney Legend.

Because The Disney Channel had scheduled the newly colorized version of the original *Shaggy Dog* (1959) to air October 18 and 25 (as well as its 1976 sequel, *The Shaggy D.A.*) on their Fall 1987 line-up, a Disney Legends award was conceived as an additional promotional push for "Shaggy Dog Month". MacMurray was selected for the promotion, not only because he was the star of the original film and friendly to the Disney Company, but because he had appeared in seven popular Disney films—so he had quite an impressive Disney track record.

His first Disney film was *The Shaggy Dog* (1959). The advertising campaign claimed that it was "[A] new kind of horror movie...Horribly Funny!" The idea for the film came from a novel, *The Hound of Florence* (set during the Renaissance), written by Felix Salten, the author of *Bambi*, and developed for the screen by Bill Walsh.

Walt initially planned to develop *The Shaggy Dog* as a television series for ABC, since the network was continually pressing him to produce more programs.

When Walt pitched the idea of a family comedy series about a boy who occasionally turns into a sheepdog, Jim Aubrey and the other ABC executives were visibly underwhelmed (Aubrey even excused himself from the meeting after Walt's pitch of the idea). Walt decided to make a feature film out of the story. It turned out to be the highest grossing film of the year.

Writer/Producer Walsh wondered if the success of the film provided a springboard for the concept of *My Three Sons* (1960-1972). "The same dog, the same kids, and Fred," Walsh remembered.

The film was the first live-action comedy made by the Disney Studio. To save costs and help hide the primitive special effects, it was made in black and white.

Although the character MacMurray plays is said to be a postman, he is never once shown working. Apparently, he was made a postman to help explain his dislike for dogs, which adds to the humor when his oldest son—thanks to an ancient curse—transforms into a sheepdog in a twist on the teenage werewolf idea.

Kevin "Moochie" Corcoran, who played MacMurray's youngest son in the film, remembered:

> I think Walt saw himself in Fred to a great extent and Walt saw something about Fred that again was a bigger version of every man. He was one of the most underestimated actors of all time.

Songwriter Richard Sherman agreed:

> Actually, Walt did identify with Fred, and especially the characters he played like the Absent-Minded Professor. That was exactly what Walt did.

The following year, MacMurray appeared on screen and received critical acclaim for his performance as the callous adulterer in Billy Wilder's *The Apartment* (1960). That same year he also appeared in the first season of *My Three Sons* (with Disney actor Tim Considine and Mouseketeer Don Grady as two of the sons). Despite his appearance in the Wilder film, most audiences still thought of him primarily as the lovable and family-friendly single father, Steve Douglas, making it easy for Disney to continue to cast him in a series of popular live-action comedies.

MacMurray said:

> When our twin girls were 5 or 6, I took them to Disneyland. While they were on the merry-go-round, a woman came up to me and said, "Oh, Mr. MacMurray, I've enjoyed your movies through the years. I saw *The Apartment* [in which MacMurray played a philandering boss] last night. How could you? You spoiled the Disney image!" And with that she hit me over the head with her purse and stormed away.

MacMurray continued:

> The first picture I made at Walt's studio was *The Shaggy Dog*. I remember, during the shooting of *The Shaggy Dog*, Walt came on the set one day and said, "You know, Fred, I kinda like what you are doing in the picture and I've got another idea for another picture after this one if you'd be available." I said, "Oh, I'll BE available!" Then I said, "What is it about, Walt?"
>
> And he said, "Well, it's just an idea I got. I was just over at the World's Fair in Brussels. There was a Doctor Julius Miller who was giving a demonstration on atomic and nuclear energy. He does it in such an amusing way I think that maybe we can do a character for you patterned after Dr. Miller and make some kind of a picture out of it. We'll see." So I went home that night and said to my wife, "You know, Walt is talking about another picture for me. I hope it comes out. I mean, I hope it comes true, but I mean it probably won't. It was just something he was talking about." But that is the way Walt worked.
>
> This is the way he did things. The next day on the set, Bill Walsh, the writer, came out and said, "Well, I've got an assignment to do a picture for you." I asked, "What's that?" And he said, "The one Walt was telling you about the professor." And that all turned out to be *The Absent-Minded Professor*, which was very successful. It kind of shows you the way how things came out of Walt's head. Amazing.

In 1961, MacMurray starred in *The Absent-Minded Professor* and repeated that role in *Son of Flubber* in 1963. In between he starred in *Bon Voyage* (1962). The last two films he did while Walt was alive were *Follow Me Boys* (1966) and *The Happiest Millionaire* (1967).

To help promote "Shaggy Dog Month" on the Disney Channel, MacMurray and his wife, actress/dancer June Haver, were invited out to the Burbank studio for what was thought to be a simple in-house celebration with a handprint and signature in a cement square inspired by the famous ceremony at the Grauman Chinese Theater forecourt.

It was CEO Michael Eisner who thought that the ceremony could be expanded into an annual event and become more than just a one-time publicity event.

Disney Legends Promenade, a section of sidewalk in front of the Studio Theater, was to contain the handprints and signatures of all honorees. Eventually, the Legends Awards outgrew the area (more than 200 men and women have since been honored) and was relocated to the newly named Legends Plaza facing the Team Disney building in Burbank, where handprints and signatures are now reproduced as bronze plaques

On Tuesday, October 13, 1987, at noon, hundreds of Disney employees and press representatives gathered to honor MacMurray as he arrived, sitting in the back of a 1915 Model T Ford reminiscent of the one he flew as Professor Ned Brainard in *The Absent-Minded Professor*

He came with his wife and a sheepdog representing the original Shaggy Dog (and in true Hollywood tradition, the dog was wearing sunglasses). In 1945 MacMurray met Haver when they co-starred in *Where Do We Go From Here?* They married in 1954 and were both at the opening of Disneyland in 1955 and introduced to television audiences by host Art Linkletter.

Michael Eisner and Frank Wells greeted MacMurray and his wife, who both sat in director's chairs in front of the decorated theater. Eisner said:

> The promenade has been established as a means to pay permanent tribute to individuals whose talents have made a significant contribution to the company's proud heritage. We chose to establish it on the studio lot to share our rich past with the employees who will be part of our company's future.

Then he explained that the Selection Committee chooses only those people whose time-honored body of work epitomizes the Disney product, and whose contributions have enhanced or secured Disney's place in entertainment history.

That first selection committee included Roy E. Disney (Disney family), Sharon Harwood (Disney University), Dave Smith (Disney Archives), Stacia Martin (Disney Film Club), Erwin Okun (Corporate Communications), Doris Smith (Corporate Affairs/Relations), Art Levitt (Corporate Projects), Arlene Ludwig (Walt Disney Pictures), Randy Bright (Walt Disney Imagineering), Jack Lindquist (Walt Disney Attractions), and Shelley Miles (Disney Consumer Products).

"Fred MacMurray is the epitome of what we hope the Disney Legends Promenade will come to represent," Eisner said.

Wells announced that Los Angeles Mayor Tom Bradley had proclaimed the day "Fred MacMurray Day" to acknowledge the Disney award. Wells handed MacMurray a framed certificate signed by the mayor.

MacMurray, who was surprised at the size of the event, thanked everyone:

> I just thought we'd come out here today, get a few pictures taken, maybe say "hello" to the dog. This is much more than I imagined.

After sharing a few memories of Walt and the studio, and accepting a commemorative plaque from Eisner (the current Disney Legends award featuring Mickey's hand holding the wand had yet to be designed), the guest of honor stepped over to a square of wet cement and knelt, pressing in his hand prints and writing his signature as the first-ever Disney Legend. Cameras flashed. The media shouted questions. There was a photo location featuring the Shaggy Dog, Goofy, and Pluto, and prizes for Disney cast member trivia contest winners Louise Helbert, Rhonda Miller, and Carol Cotter.

MacMurray said:

> You know, in talking to Walt, I remember I was quoted once when somebody had asked me: "What about Walt?" And I said you could be talking to him and it was like he wasn't listening to you. And then I went on to say that about a month or so later, he'd come up and say, "Remember that time you said so and so?" He knew everything you said, but he had a million other things about the park and the thing in Florida and all this to think about, but he'd remember everything you said to him.

It was nice that the Disney Company remembered MacMurray, as well. It was the only year that a single person was honored. The following year, all "Nine Old Men" (Les Clark, Marc Davis, Ollie Johnston, Milt Kahl, Ward Kimball , Eric Larson, John Lounsbery, Wolfgang Reitherman, and Frank Thomas), as well as Ub Iwerks, became Disney Legends.

Fred MacMurray died at the age of 83 on November 1991 of pneumonia, as a result of contracting chronic lymphocytic leukemia. He was one of Hollywood's wealthiest citizens, thanks to good investment deals in real estate, but was known as the "frugal millionaire" by carefully watching his money and doing things like bringing his own brown bag lunch to work.

The Shaggy Dog proved Disney's biggest box-office hit of the time, earning more than $8 million in its first domestic release. The studio re-released the film in 1967, and in 1976 produced a sequel, The Shaggy D.A, directed by Robert Stevenson and starring Dean Jones and Tim Conway.

Several made-for-television films followed, included 1987's The Return of the Shaggy Dog with Gary Kroeger as Wilby Daniels and 1994's The Shaggy Dog, which starred Ed Begley Jr. In 2006, Buena Vista released a remake of The Shaggy Dog directed by Brian Robbins and starring Tim Allen and Kristin Davis.

Disney and the Rose Parade

On January 1, 2013, the 124th Pasadena Tournament of Roses parade (popularly known as the Rose Parade) featured a massive float from Disneyland titled Destination: Cars Land that re-created the Radiator Springs area at Disney California Adventure in an all-floral tribute.

The float was designed to look like it breaks down several times on the parade route with white smoke spewing from it, but is always rescued when Mater the tow truck following closely behind gives it a "jump start" to get it going again. Riding on the float were Disney Channel *Shake It Up* series star Zendaya and Radio Disney's DJ Ernie D who broadcast live from the float during the parade.

As Erin Glover, Social Media Director at Disneyland, described it:

> Lightning McQueen and Sally racing in the front, on top of a floral "Welcome to Cars Land" sign; Flo's V8 Cafe and Cozy Cone Motel signs (along with a mini Cozy Cone Motel underneath); floral representations of ride vehicles from Luigi's Flying Tires and Mater's Junkyard Jamboree; Luigi and Guido cheering on miniature Radiator Springs Racers vehicles racing through a floral replica of Ornament Valley—with a real waterfall!

(The cones and signs spun around, something that is traditionally referred to as "animation" in the official parlance of the parade.)

The float marked the 75th anniversary of Disney participation in the parade. It was just another of the many Disney anniversaries celebrated in 2013.

The first Tournament of Roses Parade was staged in 1890, but Disney's participation did not begin until 1938 as a way of promoting the release of *Snow White and the Seven Dwarfs*. The film had premiered December 21, 1937, but would receive a general theatrical release on February 8, 1938.

Dancer Marjorie Belcher, who was the live-action reference model for the character of Snow White, had appeared on Christmas Eve with costumed versions of the Seven Dwarfs in the Hollywood Christmas Parade (known at the time as the Santa Claus Lane Parade) as they rode down Hollywood Boulevard. She and performers in dwarf masks (who had appeared at the film's premiere at the Carthay Circle Theater) also appeared on a Snow White-themed floral float on January 1, 1938, for the Rose Parade.

Belcher was 18 years old and, instead of the costume she used for filming, Disney got a colorfully bright sequined vest for her to wear with a fancier outfit so that it could be clearly seen by the people who crowded on the sidewalks to catch a glimpse. Roy E. Disney, Walt's nephew, who was about to turn eight years old on January 10, claimed that one of his most vivid childhood memories was watching that 1938 parade:

> I watched the whole parade standing on the hood of an old Buick in a used car lot on Colorado Boulevard that belonged to one of my dad's best friends. I couldn't have been more excited that day, especially when the *Snow White* float went by.

Roy E. Disney was the Rose Parade Grand Marshal decades later for Celebration 2000: Visions of the Future, which tied into the release of the animated feature film *Fantasia: 2000* that was set to debut January 1 at various theaters, including one in Pasadena.

The *Snow White* float was the first movie studio-sponsored float in Rose Parade history and received a "special award" from the judges, in the form of a marble electric clock. The side of the float read: "Snow White and the Seven Dwarfs—Walt Disney Productions."

The defunct newspaper, the *Los Angeles Herald & Express*, in its late edition of January 1, 1938, proclaimed:

> Snow White Brings Acclaim by Children

> Squeals of delight came from children in the great crowd as Walt Disney's *Snow White and the Seven Dwarfs* came abreast. It brought a lovely forest scene, with Little Snow White seated beneath a gargantuan toad stool, with each of the Seven Dwarfs having a smaller stool to himself, surrounding her in a rough circle.

It wasn't until 1955 that a Disney float showed up again in the parade.

The Helms Bakery float in the Rose Parade that year won the Judge's Special Award for its portrayal of the forthcoming Disneyland that would open in a little more than six months.

The float was three connected circles with Mickey Mouse at the front, Sleeping Beauty Castle in the back, and flying pink Dumbo elephants in the center circle. There were more than 7,000 pink roses on the float, which also featured a long, curved pole with a silver balloon and red lettering spelling out "Disneyland" in the center.

Helms Bakery was a very popular southern California bakery that, at one time, had bakery trucks that went through neighborhoods like ice cream trucks. Customers could purchase freshly made bread, donuts, and other baked goods, often hot from the oven. (Willie's Churros stand in Disney's California Adventure was designed and painted to resemble those fabled Helms Bakery trucks.)

The Firehouse Five Plus Two jazz band, under the direction of Disney animator Ward Kimball, also performed in the parade.

In 1960, the tradition of the annual visit of the Rose Bowl football teams to Disneyland began when the University of Washington and University of Wisconsin players toured the park prior to their January 1, 1960, game.

Of course, being the prime southern California vacation destination, it was only natural that out-of-state visitors would want to see Disneyland, but it led to more formal visits.

On December 26, 1962, Disneyland hosted the 1963 Tournament of Roses Queen and Royal Court, along with USC and University of Wisconsin coaches and players, who were to do battle on the Rose Bowl gridiron less than a week later. Mickey Mouse presented a bouquet to the queen and her court in Town Square. Then the queen, her court, the players, and the band paraded down Main Street, followed by the entire group attending a Golden Horseshoe Revue performance and dinner with Walt Disney.

That Rose Court Jubilee, as it has become called, has been an annual event, lasting for more than a half-century, even though both the Golden Horseshoe Revue and Walt Disney are both long gone.

However, when most of us think of the connection between Disney and the Rose Parade, it is the 1966 parade when Walt himself was grand marshal.

J. Randolph Richards was the president of the 77th Annual Tournament of Roses, and he reviewed more than 7,000 suggestions for a theme from the cards and letters that came in. Realizing that the wonders of the present era of 1965, such as jet plane travel, communications satellites, and orbiting space vehicles, "have seemingly reduced the size of the globe", Richards settled on the theme of "It's a Small World".

On March 14, 1965, Richards made the announcement of the theme, as well as his choice for grand marshal: Walt Disney. He felt it appropriate to so honor Walt because of:

> [T]he universal acceptance of the Disney creations. He [Walt] has penetrated barriers and boundaries, lessening the distance between the continents. This master showman has brought joy and laughter to millions in every part of the world. Many of the countries have paid tribute to his genius by singling him out for high honors.

Richards was probably also inspired by Walt's announcement that the popular "it's a small world" attraction from the New York World's Fair would be relocated to Disneyland to open in a magnificent Mary Blair-designed attraction building on May 28, 1966.

The official press announcement declared:

> Mickey Mouse [costumed performer Paul Castle] will ride in the Grand Marshal's automobile [a white Chrysler Imperial] along the

side of Walt Disney, whose agile pen created the lovable animated character many years ago. The rest of the Disney characters will be close by. A total of 27 of the widely known cartoon creations will be taking part in the New Year's Day Festivities by walking along behind the automobile of the famed showman or riding on the entry of the city of Burbank, a design of [the] Disney Studio in that community.

In January 1988, Paul Castle told a reporter for the *Los Angeles Times*:

My most favorite time of all [portraying Mickey Mouse] was with Walt Disney in the Rose Parade in 1966, the year he passed away. He was the grand marshal of the Rose Parade and I was in the car with him, in the back seat, just Walt and I for three hours. Just Walt and I. Of all the things I've done in my lifetime, that to me was my biggest day. Walt and me. January 1, 1966.

The official press release stated:

It was Walt Disney's choice that if there was to be an entry revolving around his life and his creations that the Burbank float should be the one to carry out the motif. The Walt Disney Studios in that city is closely tied in with the story of this man and his creations. Therefore, in all probability such a choice was a sentimental one, reflecting his esteem for the San Fernando Valley municipality.

Ironically, Burbank had denied Walt permission to build Disneyland there, fearful of the carnival atmosphere it might generate along with a "bad element" of visitors that often frequented such venues.

The Burbank float, with design help from Disney Legend Bill Justice, was titled "Our Small World of Make Believe", and was the only float that year featuring Disney characters.

Walt started the parade riding in an open car followed by the Burbank float which was described as "an open book, a musical clef and an artist's palette, representing three important elements of the Disney legend—the story, the music, and the creative art work".

Costumed Disney characters who appeared on and around the float included Winnie the Pooh, Captain Hook and Peter Pan, Pinocchio, Gideon, J. Worthington Foulfellow, Pluto, Goofy, Alice, the White Rabbit, the Three Little Pigs, Big Bad Wolf, and the Seven Dwarfs.

Artists at the Disney Studio contributed artwork throughout both the official program and the participant dinner program, using many of the standard Disney characters, "it's a small world" attraction designs, and Disney-drawn mascots for the teams of the two conferences, including the notorious Oregon Duck who bore more than a passing resemblance to Donald Duck.

Walt and his wife, Lillian, attended the actual Rose Bowl game, seated in the box of the tournament president, where the UCLA Bruins defeated the No. 1-ranked Michigan State Spartans by a score of 14–12. The 150-piece Michigan State Marching Band performed earlier at Disneyland on December 30, 1965, as part of a special Michigan State Day.

The Purdue University football team visited Disneyland on the afternoon of December 22, 1966, exactly one week after Walt Disney died. The Purdue Marching Band paraded in Disneyland, along with the USC Trojan Band, on December 30, 1966. Indiana University and USC's football teams visited the park on the afternoon of December 22, 1967. The Indiana University Band paraded down Main Street at Disneyland on December 29, 1967.

For the 1971 parade, Disney worked with Anaheim as consultants and designers on their 1971 Tournament of Roses Float entry. The title of the float: A Dream Come True in Anaheim.

The official description of the float was:

> Anaheim creates, in flowers, the dream that came true within the city's boundaries, of course, this dream come true is DISNEYLAND. Appearing on the entry arc 22 of the famous Disney characters— Mickey Mouse, Pluto, Goofy, Show White, Alice in Wonderland, The Wolf and Three Pigs. Some of these are highly animated floral figures. At the rear of the float are the newest characters featured in the latest Disney full-length cartoon feature, The Aristocats. Two "youngsters", Angela Dutton and Jimmy Sundali, appear on the float in a big over-sized bed. Also on the float is a floral version of the castle and moat.

Disney gave permission to the Sunkist Corporation to use characters from Song of the South on their Tales of the Briar Patch float that same year. Disney had released a statement in the entertainment newspaper Variety on February 25, 1970, that the film was "permanently on the shelf as offensive to Negroes and present concepts of race", but re-released it in 1972, where it became the highest grossing Disney re-release up to that time.

Appearing on the float were characters from the film including Brer Rabbit, Brer Fox, Brer Bear, and other animal and bird figures, such as crows and squirrels. The official description of the float was:

> Brer Fox peers forth from his habitat in an old log, while Brer Bear stands in the middle of a winding path. Hanging from a branch just behind Brer Bear is Brer Rabbit. Two black crows are surveying the scene from their perch high in the brightly colored autumn trees.

Rose Bowl Football teams from Ohio State and Stanford met for the first time on December 21, 1970, at Disneyland. The teams visited the park with their coaches, Rose Queen Kathleen Arnett and her court, along with their hostess, Disneyland Ambassador Marva Dickson. After a greeting at the front gate by the Disneyland Band and Mickey Mouse, the group marched down Main Street. Goofy and the Scatcats, from The Aristocats, joined the group in front of Sleeping Beauty Castle for photos.

In 1973 the theme of the parade was Movie Memories, so Disney used it as an occasion to tie in with its 50 Happy Years promotion. (The Disney Bros Studio began in 1923 producing the Alice Comedies).

The Disney float started the parade (directly after the motorcycle riders), but it was actually a caravan extravaganza, including four Castle turrets, a Castle float, the Love Bug, a Wishing Well, Monstro the Whale, three Teapots, a Circus Train, a Hunny Pot, a Floral Plaque, two Peter Pan-themed pirate ships, and an Uncle Scrooge cycle. The official description was:

> This unique and special entry in the 1973 Rose Parade honors the 50th Anniversary of Walt Disney Productions. A block-long cavalcade of more than 100 famous Disney cartoon characters promenade in a fanciful atmosphere of make-believe to recreate the most celebrated memories from Walt Disney's film classics.
>
> Movies represented are *Cinderella, Snow White, Mary Poppins, Alice in Wonderland, Jungle Book, Song of the South, Dumbo, Pinocchio, Sleeping Beauty, Winnie the Pooh, Peter Pan, Bedknobs and Broomsticks, The Love Bug, The Aristocats,* and other famous Disney Characters.

Characters not in the feature films included Mickey, Minnie, Pluto, Goofy, Donald, Horace Horsecollar, Clarabelle Cow, the Three Little Pigs and Big Bad Wolf, Chip 'n' Dale, and Scrooge McDuck.

The City of Glendale decided to produce a float to join the celebration of Donald Duck's 50th anniversary that was still ongoing for the 1985 parade. To further enhance the float, they included Clarence "Ducky" Nash, a longtime resident of Glendale and the decades-long voice of Donald Duck.

Unfortunately, Nash became ill and was unable to ride in the parade as planned, after he was taken to St. Joseph Medical Center on New Year's Eve. Tony Anselmo, the current voice of Donald, had gone to the parade specifically to see Nash on the float and was disappointed. The beloved Nash died the following February.

On March 11, 1985, the Pasadena Tournament of Roses gave a special certificate of appreciation to Disneyland "[I]n recognition of your contribution to the success of the Tournament of Roses and with thanks for many years of support".

In 1986, Lawry's licensed Mickey and Minnie for their float and produced a pin, but, because they had mistakenly not gotten permission for any merchandising, only 14 pins of Mickey and Minnie waving from the back of the float were produced and given to the Lawry's board of directors. The official Tournament of Roses pin guide for that year lists the pin as "not available".

For the 1988 parade, four Disneyland Electrical Parade float drivers were chosen to drive the California Bicentennial Foundation Float (celebrating the 200th anniversary of the signing of the Constitution). The Disney Company was a co-sponsor and the float featured an original

Disney-designed buffalo character, Bicentennial Ben, the official mascot of the foundation.

Mickey Mouse rode on top of the We Are the People float, the first float in Rose Parade history to represent an entire state, California, and depicting "the Constitution and Independence Hall, connected by two giant flags to a future represented by astronaut Buzz Aldrin astride a Martian landscape".

Disneyland drivers were also at the controls of the Pacific Financial Company's float and the Pac 10 float.

In December 1992, Disneyland hosted another Rose Court Jubilee featuring Tournament of Roses Queen Liana Yamasaki and her court, along with bands from participating college football teams in a parade down Main Street. Disney helped design the City of Glendale's 1993 float that featured Sorcerer Mickey.

Jack Lindquist, then president of Disneyland, told the Los Angeles Times in its November 12, 1992, edition that the Disney characters would be contributing "a new dimension in this great New Year's Day celebration". Lindquist said the "Mickey's Toontown" characters will present a pageant "saluting the [1993] tournament's theme", which is Entertainment on Parade, during the march down Pasadena streets. He said Disneyland also would sponsor the 1993 parade's theme float, a multi-level Toontown house, to tie-in with the opening of Mickey's Toontown.

On December 29, 1993, the marching bands from UCLA and Wisconsin performed on Main Street, U.S.A. in the Rose Court Jubilee, a special parade that was featured during a one hour CBS TV special, Coming Up Roses, which aired on January 2, 1994. This time the colorful pageant featured the princesses or heroines of Disney animated classics accompanying Rose Court Queen Erica Beth Brynes and her court.

In 1995, Disneyland sponsored a float in the parade titled Salute to Sports, featuring floral statues of Mickey, Minnie, Donald, Goofy, and Pluto in sporting gear, and live cheerleaders in front of Sleeping Beauty Castle. Disneyland again hosted the Rose Court Jubilee parade in December 1994 with the Rose Queen and her court and the football teams.

On December 29, 1996, there was a Rose Court Jubilee parade at Disneyland, along with bands from Northwestern and USC, but this time there was a "battle of the bands" with Nortwestern in the Castle forecourt and USC in Town Square.

For the 1999 parade, Disney assisted with the entry from the Los Angeles City Department of Recreation and Parks WOW (Wonderful Outdoor World) program. Eight 12-year-olds were asked to create an equestrian unit, but all of the children were brand-new to riding. Walt Disney Special Events helped with the creative concept for the unit, including costumes for the kids and decorations for the horses and mule train.

Over the years, Walt Disney Imagineers and Disney Entertainment executives have served as judges in the parade. As mentioned, Roy E. Disney was the grand marshal for the 2000 parade, making him the first member of a family of a previous grand marshal (Walt) to be so honored.

For the first time in Rose Parade history, the parade began not with a traditional theme banner or a marching band or a float made of flowers, but with a "human theme banner" comprised entirely of people. Designed by Stadium Stunts, it featured 2000 colorfully costumed southern California high school students creating a formation spelling out the theme of the parade: Celebration 2000. The shape and color of the unit changed on musical cue to read Fantasia 2000, the title of Disney's newest animated feature film, which opened in IMAX theaters on January 1, 2000.

On December 27, 1999, there had been another Rose Court Jubilee at Disneyland. Tournament of Roses' President Kenneth H. Burrows and Rose Queen Sophia Bush and Disney characters rode cars and floats down Main Street in the parade.

In 2003, members of the Disney VoluntEars from the studio helped to decorate the Make-A-Wish Foundation Tournament of Roses Parade float.

For the 2004 parade, to promote the opening of the new Disney's California Adventure attraction Twilight Zone Tower of Terror, Disney produced a float called A Sudden Drop in Pitch. The musical theme from *The Twilight Zone* television show played on the float over and over in keeping with the parade theme of Music, Music, Music.

The float was the tallest in Rose Parade history, with a height of nearly 110 feet. It was so tall it had to be designed with 11 hydraulic pumps to fold it down to 17 feet when the parade went under freeway overpasses. The tower was placed in the midst of Disney California Adventure landmarks like a miniature moving version of the Sun Wheel.

There were four stunt people on the tower that shook when "lightning" struck it. The exposed elevator shook as if caught in a tremor, and the roof sparked with flashes of lightning. In addition to the carbon dioxide smoke effects, the float was the second in parade history to use pyrotechnics.

There was some minor controversy that relatives of victims of the 9/11 tragedy felt that a tower that would fall and had the name "terror" was disrespectful. There was also some criticism that Disney was more interested in commercially advertising its attraction than sticking to the actual theme of the parade.

The Rose Court Jubilee was held December 26, 2003, at Disneyland.

In 2005, to promote the 50th anniversary of Disneyland, Disney's float was called The Happiest Celebration on Earth.

For the first time in Tournament of Roses Parade history, a two-and-a-half minute opening ceremony and show featuring 150 Disneyland cast members was performed, with the Disneyland float as its backdrop. After the ceremony, the float, decorated by cast members, led the parade.

The airy float, which drew its inspiration from Disney park castles with multiple turrets and spires, also featured members of the worldwide ambassador team, one from each Disney theme park, and a variety of Disney characters. Mickey Mouse was the grand marshal for the parade.

Tournament of Roses President Dave Davis said:

> Mickey Mouse has brought entertainment, joy, and laughter to families around the world for 75 years, and we couldn't think of a more ideal Grand Marshal to help us "Celebrate Family" in 2005. Mickey Mouse became a part of the Tournament of Roses family when he accompanied Walt Disney on his Grand Marshal ride in the 1966 Rose Parade, and we are delighted to welcome him back once again to help us spread New Year's cheer on January 1, 2005.

In 2006, Disney had three floats in the Rose Parade. The first, titled The Most Magical Celebration on Earth, wanted to remind the audience that the 50th anniversary celebration was still going on. The float, at 150 feet, was the parade's longest, and included five towering, strawberry powder-covered castles representing Disneyland, Walt Disney World, Tokyo Disneyland, Disneyland Paris, and Hong Kong Disneyland. Each castle was surrounded by plants indigenous to their regions of the world, since the Disney Company was emphasizing that it was not just Disneyland's 50th birthday, but rather the anniversary of the birth of Disney theme parks worldwide.

There was also a Little Einsteins float. Made of flowers, seeds, bark, leaves, and other natural materials, the animated characters of Leo, Annie, Quincy, June, and Rocket were interactive, with Rocket soaring 25 feet in the air. The float, titled Making Music is Magical, was a cooperative effort between the Disney Company, the Baby Einstein Company, and the International Music Products Association.

So the Disney Company has had a long and colorful connection to Pasadena's famous Rose Parade that will continue for many years to come. In fact, the use of color in Disney parks' topiaries was inspired by the iconic use of floral color on the floats.

.

Disney Animator Pranks and Hijinks

Right from the beginning, Walt Disney encouraged a humorous climate at the Disney Studio to alleviate the constant stress. As Ward Kimball told me in 1996:

> There was an element of boredom when you're animating. You're doing all these characters that move ever so slightly from one drawing to the next and it gets so repetitive that you think of excuses to take a break and blow off some steam.

In a 1940 *Atlantic Monthly* magazine article about the Disney Studio, Paul Hollister described it as "the only factory on earth where practical jokes are part of the production line".

Richard Greene, in his book *Man Behind the Magic*, wrote:

> Walt didn't join in on the hijinks, but he was tolerant of them. As long as good work was being turned out, he would put up with almost anything

> On one occasion, Walt Kelly—a Disney animator who went on to create the comic strip *Pogo*—targeted a fellow animator who took great pride in successfully throwing his coat across the room onto a coat rack. Kelly sawed the coat rack into dozens of small pieces and taped it back together so the breaks wouldn't show. When the victim came back from lunch, he tossed his coat, as usual, and nearly passed out when the whole rack came tumbling apart like a house of cards.

The victim of that prank was Fred Moore, who had just returned from a "drinking lunch", and his friends Kelly and Ward Kimball delighted in his shock and surprise.

Animator Jack Kinney said that "the victims of these so-called jokes always had a standard comeback: 'Why don't you guys put them funny gags in the pictures?'"

Kimball was perhaps the "King of the Disney Pranksters" During my 1996 interview with him while I was working at the Disney Institute in Florida, I prodded him to share some of his memorable pranks.

Jim Korkis: You have quite the reputation for pranks at the Disney Studio.

Ward Kimball: When we moved into the new Burbank studio, there were very few bathroom stalls that were operating. We had a main hallway in the building and these units teeing off from it and in each hallway was a man's "can" as we called it, but only two stalls. So I got this idea. There was always a traffic jam in the morning due to not enough restroom stalls. So, one day I went down to a Salvation Army thrift store and bought 12 pair of shoes and some second-hand pants and took some wooden doweling to support the pants and shoes.

I got to the studio very early one morning before anyone had come to work. I rigged up all of these in the stalls...even the women's stalls...with these shoes on the floor and the wood supporting these pants that I had pulled down and set on all these thrones and locked the stall doors. Then I went to sleep at my desk where an hour or two later I was awakened by people pounding on the stall doors and yelling. All hell broke loose. "Give me a chance. What are you doing? What's taking so long?" Apparently, they looked under the stalls and saw the shoes and pants rumpled up, but it never occurred to any of them to look over the top. They'd all look down under. Eventually, the gag was discovered.

JK: Another version I heard involved the use of cels.

WK: That was another time. We took some cel material. Remember, it was transparent. And we covered the top of the toilet bowl with it and then put down the lid. The women never suspected when they sat down to use the facilities until it was too late.

JK:. Let's just say that your practical jokes were legendary, like at Ben Sharpsteen's wedding and the wrap party for *Snow White and the Seven Dwarfs.*

WK: Not everyone cared for Ben Sharpsteen. A lot of the guys felt he was Walt's "hatchet man" and he could be pretty hard, you know. So I knew they'd get a laugh at him getting back a little of his own, if you know what I mean. For Sharpsteen's wedding, I hired a life drawing model to walk down the aisle completely naked except for a wedding veil and holding a baby to disrupt the ceremony.

At one of the wrap parties for *Snow White*, I hired a guy to dress up as a policeman and come in to the party and harass Walt that the party was too noisy and he would have to take them all in to jail. Well, by the time the guy finally showed up, Walt was so drunk himself that he kept arguing with the policeman and telling him he was going to have his badge.

Another prank animators used to play on Sharpsteen that he never knew about was when the animators were "dipping their pen in company ink", the term used for having sexual relations with the ink and paint girls, they would sign Sharpsteen's name in the motel registers rather than a standard "John Smith".

In 1997, I asked Disney animator Bill Justice, famous for his work on Chip 'n' Dale, to share a few of his memories about pranks at the Disney Studio.

Jim Korkis: I know that Disney animators would break up the tedium doing a lot of pranks. Were you involved in any of those?

Bill Justice: Pranks? Some of it was pretty standard. We'd balance a cup of water over a door to drench someone when they opened the door. One time someone put a roll of animation paper up there. That must have hurt when it fell and hit the guy.

One of the gags was taking the animation discs off and putting a kneaded eraser or a piece of limburger cheese on the incandescent bulb so that it would burn and stink under the disc when an animator was drawing. They didn't know where that terrible smell was coming from because it was so gradual.

In the old Hyperion studio, there was a little garden area with some benches and people would go on break. Walt would go there. There was a little turtle that would eat the vegetation. So the guys got the idea to bring in a different turtle. They kept bringing in a larger and larger turtle every two or three days so it looked like the turtle was growing at a fantastic rate. Then, to top the gag, they reversed it and brought in smaller turtles every two or three days. Walt even started talking about it at meetings. I am sure he caught on to the gag, but in the beginning I think they got him going.

When I talked with Ken Anderson in 1985, he told me a legendary prank on artist and storyman Roy Williams at the Hyperion studio. Because of his size (320 pounds and a former All-American high school football player) and nature (a tremendous temper and the mouth of a sailor when it came to swearing), Williams was often the victim of gags.

Storyman Dick Kinney would have a squirt gun under a table and shoot Williams in the crotch while they were talking over a story for an animated short. Since Williams had a huge stomach, when he stood up he couldn't see the area beneath his belly and would walk around while others would secretly laugh as he went by, thinking that he had peed in his pants again.

Jim Korkis: Everyone I talk to has a different Roy Williams story and they are all terrific. Do you have one?

Ken Anderson: He was a naïve child in this great body who could throw people around. An enormous gag man. He just churned out these cartoons like you wouldn't believe. Walt decided that Roy should be a little more dignified, so we helped Walt out. When Roy was made a gag captain at the old Hyperion studio, we made it an important thing. He had to wear a suit, tie, and a vest...and socks. Everything. Then he came over to this new building. It was really just a couple of apartments we had gutted and made into rooms for the storymen.

We had Roy all dressed up. I got Ethel, who was his wife, to make sure Roy wore the suit and everything. It didn't really fit. Nothing buttoned. Nothing really worked. As gag captain, one of his jobs was to go around and "pass" on the gags that everyone had done. He came into the room. We took this little guy, Joe Sable, who was a new guy who was maybe 5 foot 2 and weighed all of 80 pounds.

We took [storyman] T. Hee's pants. T. Hee was a big man in those days. He lost over 300 pounds. We took his pants and wrapped them around and around Sable and tied it up with a belt. So Joe is saying, "Mr. Williams, is this gag acceptable?"

And Roy is going crazy. "Are those your pants? For crying out loud, are those your pants? Have you been on a diet?" Joe says, "Yes, sir, but what's good for one person isn't necessarily good for another." Roy is getting desperate. "Never mind that. What have you been on?" Joe says, "I hate to tell you, but it was sauerkraut juice. You should probably check with your doctor. Now about this gag..."

Roy bolts away and calls his doctor. At least he had that much sense. He got the nurse and asked her if sauerkraut juice is good for diets and she says, "Yes, but..." and Roy hangs up before she finishes and runs across the street and gets a gallon of sauerkraut juice and drinks this whole can. The meeting with Walt on gags is due to come up in less than 10 minutes. This whole business began to work on poor Roy's insides and there was a lot of Roy to work on.

Have you ever heard elephants trumpet? That was the sound coming out of him. Roy would come in to the boards and these sounds are starting. Rumble. Rumble. Boom. He runs down the hall to where we had a lavatory and we hear Bang! Bang! Bang!

We had the doors all locked. Then we had the next building fixed the same way. Wherever he went there was no chance for him to get any relief. Walt comes in and sits down on a camp chair and he is already drumming his fingers and saying, "All right. All right. What have you boys got here?"

Roy tries to start telling Walt the gags and he just can't take it any more. Roy bolts out of the door and knocks over everyone in his way and makes it across the street to the main building finally.

The most classic prank story at the Disney Studio has been told to me by several different Golden Age Disney artists, with a few variations. Here is how artist Floyd Gottfredson told this prankish tale:

> The whole thing happened in the Comic Strip Department and the principal characters were [writer/artist] Ted Thwaites and [artist] Al Taliaferro [who did the Donald Duck comic strip for decades].
>
> We all worked in the back room of the annex at Hyperion [studio]. Ted carried his lunch in a brown bag and every day brought in a small can of fruit cocktail and he loved it so much and he smacked his lips over it and he'd tell Al, "I just couldn't eat a lunch without this."
>
> So this started Al's brain to working and one day he brought in a can the same size, a can of mixed vegetables. When Ted went out of the room, he would always tell Al where he was going. So, the minute he got out of sight, Al would jump up and take the label off and put rubber cement on the thing and wait until it almost dries—and just switch the labels from the mixed vegetable can to the fruit cocktail.
>
> So Ted came back the first time and he opened this thing and he actually took a spoonful of the stuff before he noticed it was not his fruit cocktail. Al, of course, was watching him.
>
> Ted stopped—then he took another spoonful of the stuff and he looked at it and he says, "I can't believe this!" He was very British and very gullible. He says, "Something's wrong here."
>
> So he shows it to Al and Al peers at it and says, "What's wrong? What is that---vegetables?"
>
> Ted says, "Yeah! Look at the label—this is supposed to be fruit cocktail."
>
> Al says, "That's strange."
>
> So between the two of them they decided that some of the labels had gotten mixed up at the canning factory. There wasn't any more said about it, except Ted went around and told everyday in the department. He couldn't get over it.
>
> Al let it go for three or four days and then switched labels again. Ted said, "The only way I can explain this is that they must have mixed up a whole lot shipment. Just imagine! These things are on the shelves of markets all over the country."
>
> Al did it just far enough apart to keep Ted intrigued. He'd have peas or carrots and even hominy one time—and Ted had never seen hominy before. To put a little variety in the act, Al reversed the procedure and put a vegetable label on the fruit cocktail can.
>
> "That's crazy!" Ted says, "I know I bought fruit cocktail this morning. Al, look at this!"
>
> Al says, "What's wrong with it?"

Ted says, "That's mixed vegetables!"

Al says, "That's funny. You must have picked it up by mistake." So Ted opens the can and it has fruit cocktail in it.

Finally, it was Ted himself who said, "Well, I really think this is an item for Robert Ripley's Believe It or Not. I think I should write it in to him and maybe I'll get some money out of it."

We all agreed, and by this time, everybody knew about it. Ted actually wrote Ripley. After he had written to Ripley, we knew we had to do something to pay this whole thing off. We wondered for two or three days what we could do. We figured it would take eight days before he would expect an answer.

The plan called for this last can to be mailed to Thwaites, with appropriate King Features labels (the newspaper syndicate that distributed the Disney comic strips but also *Ripley's Believe It or Not* comic panel) made up by the comic strip men, and was to contain a rather potent message from Mr. Ripley inside the can along the lines of "I don't believe it! (signed) Robert Ripley."

But Al couldn't leave it alone. He had to switch one more can and Ted came back too soon and Al had to rush it.

When Ted came in and ate his lunch right after that, he picked up this can and the label slipped off the can and here was this wet rubber cement. He stopped and looked at it for a minute, then he says in his British accent, "You so and so's. Suddenly everything is clear to me. I know what's been going on here!"

In some variations, Ted did open the can and discover the note. In other variations it was canned peaches. Whatever the actual story, it became the classic story of a Disney Studio prank.

Ward Kimball kept a detailed diary from his time at the studio.

In 1941, Sterling Silliphant, then the publicity director at the Disney Studio, but who went on to fame and fortune as a top screenwriter on projects like the television series *Route 66*, contacted Kimball and his friend, legendary animator Fred Moore, to do local Saturday kiddie matinees at two different theaters and draw some Disney characters.

Kimball and Moore talked their friend, animator Walt Kelly, who would later go on to create the comic strip *Pogo*, to be the emcee. Kelly agreed, but only if he could get a few drinks in him to overcome his shyness.

Here is an excerpt from Kimball's diary, March 22, 1941, of the results:

Kelly, Fred, and Kimball meet at 11:30 a.m. at the Blue Evening for bourbon and sodas. We could hardly navigate when we left for the Stadium Theater on West Pico for the first show. Met Silliphant and theater manager.

We went on after a Donald Duck short. Lots of kids in the house. Kelly droned on and on with unrehearsed double-talk as Fred and I drew Disney characters. Kelly gave us a lot of asides which broke us up.

Kelly, at one point, said to the audience, "I suppose you are wondering why Ward is wearing a railroad conductor's hat. Well, he's got a bald spot on the back of his head, and we told him to put black paint on it, but he refused and wore a hat instead." The kids clapped and liked our act.

We did the same routine again at the Fairfax Theater, to a better reaction. We were more sure of ourselves. The lights were hot and I was sweating. The kids would yell things like "Draw Donald the Duck! Draw Snow White!"

We told Kelly that if the little bastards wanted us to draw Snow White, steer them away, because she was hard to draw. So when they'd yell for Snow White, Kelly would turn to us and say, "Gentlemen, we have another request for you to draw Donald Duck." We got to laughing so hard that we would break our chalk, and I guess it became pretty obvious we were drunk.

Everything seemed to be going fine until a small bottle of gin fell out of Fred's pocket when he was rummaging around for a piece of chalk. This brought the act to a quick close, and I remember at the time, during the ensuing excitement, the manager running out on the stage and saying, "You boys have got to get off! This is no good for the youth of America!" Throughout it all, Kelly stuck to the mike telling Irish jokes.

One of the popular activities during the late World War II years for the Disney animators was to shoot houseflies out of the air with rubber bands. This activity took skill, patience, and a steady hand.

Influenced by the military aviators fighting the good fight, the animators created their own tiny "fly" symbols and would put them onto the sides of their Moviolas to indicate their "kills", just as fighter pilots would add a symbol to the side of their planes.

Paper cups filled with water were pinned to the backs of chairs so that when an animator leaned back, the cup tipped up and water poured down the back of his neck. There would be races down the hallway rolling film cans. When it was raining, the noontime sports activity would reconvene inside the hallways where flying footballs might knock over lamps and other objects.

"Ward Kimball was the sort of entrepreneur of these types of activities," remembered animator and production designer Iwao Takamoto, who recalled that the killing of flies with rubber bands eventually escalated into rubber band shootouts in the halls.

Takamoto got his start in animation at the Disney Studio when he was
barely 20 years old after his release from a California internment camp for
Japanese Americans. There, he met several artists who gave him some art
training. One of them, a former art director, recommended that Takamoto
look for work at the Disney Studio because "it was a liberal place when it
came to hiring", meaning that he would not be judged for being Japanese.

In a 1995 interview, Takamoto said:

> I think that's what made the Disney Studio so successful. There was
> no prejudice about race. Everyone who worked there took people at
> face value as to whether they could do the job.

When he joined the studio, Takamoto said that he "had no idea what
animation was about. I learned on-the-job, picking up a little from some
of the great animators like Milt Kahl, Ward Kimball, and Frank [Thomas]
and Ollie [Johnston]."

He was so well-liked that, of course, he became the victim of a Disney
animator prank.

Just before he turned 21 years old, several animators invited him to
join them for a drink to celebrate the upcoming event. They went to a local
bar and started downing martinis.

Takamoto recalled:

> They asked me to have a martini with them and I told them I didn't
> know anything about martinis. I learned that martinis have olives
> and the olives are kept in a jar with a sort of brine juice from the
> olives. What I didn't know was that they had poured that juice from
> the olives into a martini glass and chilled it.
>
> They handed me the glass and said, "Here's your first martini!" I drank
> it and it tasted sour and bad, but I tried to appear sophisticated and
> cool. I said, "Nice." Then they all laughed and let me in on the gag and
> got me a real martini for my birthday that was much better.

Takamoto even got involved with a few pranks himself.

He and fellow animator Stan Green played a prank on Marc Davis,
known for many achievements, including his work on Disney villainesses
Maleficent and Cruella De Vil.

Davis was sitting at his desk in deep concentration on some project
that was giving him trouble. Takamoto and Green took up positions on
opposite sides of Davis' open doorway that led to a hall. Green began the
prank by walking in place, slowly increasing the sound as if someone were
walking toward the door. Davis heard the sound and waited to see who
was walking down the hall when they passed by his open doorway. Green
stopped and Takamoto took up the same rhythm on the other side of the
open doorway so it sounded like the footsteps were walking away.

Davis was dumbfounded. He hadn't seen anyone pass by. Eventually, his curiosity got the better of him and he went to the door to look, but Green and Takamoto had heard him get up from his chair and so they hid. When Davis got to the doorway, he looked both ways, but saw no one in the hall and puzzled over the "ghost" he had heard.

Takamoto also remembered another prank played on a newly hired animator when someone brought in an attachment meant to make Christmas lights blink on and off:

> Back then, we had to use a standard light bulb instead of a fluorescent tube under the light board of an animation desk. As luck would have it, we had a new guy working on his light board that day, so they hooked up the blinker attachment without him knowing. He pulled the switch and the light went on. He started to draw and the light went off. He just waited patiently. The light went back on and he continued to draw.

> The light went off and he stopped again. It got to the point where he actually timed the thing. Every time the light went on, he hurried like crazy and would draw as fast as he could, expecting that the light would go off...which it did. But he'd wait until it came back on, so he could go back to drawing. To watch him scramble like that... his reaction was great.

Takamoto saw no harm in these pranks:

> Today, animation is a serious business. Back then, it was a business that was, in itself, a cartoon. It was a constant series of people acting like kids.

> I thought it was cool to work with people like that. They were my role models. They could be like that, but also be real sophisticated. A lot of them were always dressed in the "high style" of the Hollywood elite.

Disney director Jack Kinney, best know for his work on the Goofy shorts, had a funny story about pranking storyman Homer Brightman, who considered himself something of a comedic actor.

Brightman's comedic storyboard pitches always brought laughter from his audience, once prompting Walt to lean over to a secretary and ask, "Are you laughing at the story or at Homer's performance?" "Homer," she replied

In 1988, Kinney wrote about a prank at Brightman's expense:

> One hot, quiet night, for want of something better to do, we started a rumor, with Homer Brightman as the patsy. We told him that Walt wasn't going to have time to do Mickey on the radio [for the *Mickey Mouse Theater of the Air* program, which turned out to be correct] and was looking for a substitute.

> Homer fell for it and went around all the next day practicing the high falsetto: "Hello, Minnie. Hi Pluto. Heh, heh, heh!"

> We convinced Homer he was a natural and set up an audition starting at 7:30 p.m. The mic was turned on and the audition began...with the entire Story Department hiding out upstairs in the next building, catching the act through the windows.
>
> Stuart [Buchanan, the person in charge of casting voices] was in the booth. After each reading, he would emerge and offer suggestions like "That was fine, Homer, but we need more action in the reading, so could you hop up and down when you read the lines? Okay, take 23..."
>
> Homer hopped. Stuart would say, "Homer, you're out of mic range, would you hold the mic as you jump? Take 37..."
>
> By 10 p.m., Homer was exhausted, sweating and pooped, but still game. Hop, hop, hop. "Hello, Minnie. Hi, Pluto. Heh, heh, heh!"
>
> Stuart came out again. "Hold it, Homer. Now your socks squeak." So Homer is struggling to pull off his shoes and socks and Stuart says, "We'll try it again when we have more time."

Disney storyman Ted Sears once visited Knott's Berry Farm, but was appalled that the Old West saloon did not serve anything alcoholic. Sears was still peeved when his wife insisted they make a reservation at the restaurant for them and their two friends. Grumbling, Sears made the reservation and was told there would be a short wait of 20 minutes for the famous fried chicken and boysenberry pie.

Sears decided to get his revenge by writing a phony name to be called. Roughly 20 minutes later, over the loudspeaker, the name was repeated again and again. The name that Sears left? "Byrdchitte", which sounds different when said aloud.

At the Disney Studio, Sears was thumbing through a phone directory and came across a name he fell in love with immediately. He dialed up the person:

"Hello? Is this Gisella Werberserk Piffl?"

"Yes, it is. How can I help you?"

"I'm an old friend of your brother's. We were classmates at Cornell."

"Oh, I'm afraid you have a wrong number. My brother did not go to Cornell. He graduated from Princeton."

"I'm so sorry," replied an extremely apologetic Sears. "You must be some other Gisella Werbersek Piffl."

Some pranks were not newly created gems, but repeats of childhood ideas. During work on the animated feature *Pinocchio*, many late nights were spent at the Disney Studio. One of the assistant directors was Lou Debney, who was called "Whitey" because he always wore white pants.

One late night, he collapsed into a nearby chair for a quick nap and one of the young animators put warm water in a film can, slid it toward Debney's arm that was hanging over the side of the chair, and gently put Debney's hand into the water. Shortly afterward, the expected result that most school children know would happen was very noticeable on his white pants.

With the death of Walt and a more corporate approach to the running of the Disney Studio, these hijinks soon faded, but did not die out completely. The pranks were a reflection of the youthful energy of the Disney animators and an effort to release the stress and anxiety of creating classic masterpieces.

Walt Disney and New Orleans

New Orleans Square in Disneyland was the first new "land" added to the theme park since it opened in 1955, but Walt had always wanted there to be a New Orleans-themed section in Disneyland from the start.

At the opening day ceremonies, co-hosts Ronald Reagan and Bob Cummings both referred to the New Orleans flavor at the edge of Frontierland, and the famous jazz band the Firehouse Five Plus Two played Dixieland jazz to inaugurate the area. A Disneyland postcard of the area from 1956 stated: "Down on New Orleans Street over in Frontierland... finest barbeque this side of the Mississippi..."

Walt planned that the Magnolia Park just around the corner from Adventureland's Jungle Cruise would eventually develop into a New Orleans area. Wrought-iron balconies, like those iconic of New Orleans architecture, decorated the exterior of Aunt Jemima's Pancake House.

The Plantation House restaurant (often referred to as the Chicken Plantation Restaurant because it featured full chicken dinners for $1.65) was designed so that the side of the building that faced the Rivers of America had a New Orleans-style façade. It was reminiscent of pre-Civil War New Orleans. The side of the building that faced the Santa Fe & Disneyland Railroad was designed to look like a Spanish hacienda, in keeping with the Frontierland theme.

In the late 1950s, the Imagineers conceived of this area as having a creepy haunted house, a pirate wax museum that guests would walk through to see tableaux of pirate history, and a Thieves Market for shopping.

By 1961, in order to put in as much as possible into the area, the haunted house was moved to the north, taking over the land where the Plantation House restaurant originally stood.

The New Orleans area was proposed to be the Blue Bayou Mart, an enclosed section where it was always a breezy summer night with stars in the sky. There would still be a Thieves Market with the pirate museum underneath the street in a "basement". However, there would now be an elegant restaurant overlooking the Blue Bayou.

Imagineer Sam McKim created a concept map. Because of the demands of the projects for the 1964–65 New York World's Fair, Walt pulled all the Imagineers off the New Orleans project to concentrate on the pavilions for the fair. When the fair ended, everyone had a new perspective, including changing the walk-through pirate wax museum into a boat ride like "it's a small world" and changing the static figures into Audio-Animatronics.

Why was it called New Orleans Square, since it doesn't seem to be a square at all, but a series of curved, winding streets? The Vieux Carré is the historic name for the actual New Orleans French Quarter, and translates from the original French into "Old Square".

The area at Disneyland was officially dedicated on July 24, 1966, by Walt and Victor Schiro, who served as mayor of New Orleans from 1961–1969.

A reporter for a New Orleans newspaper wrote that "it's the next best thing to being there" and repeated the information from the Disney publicity material that it was built for almost the exact amount paid for the entire Louisiana Purchase in 1803, roughly $15 million (just $2 million shy of the cost to build the entire Disneyland park in 1955).

When Schiro repeated to the gathered reporters that the Disney version was "just like the real thing", a playful Walt off to the side and in a soft voice said "only cleaner". Walt was in an exuberant mood and kept interrupting the mayor, pointing out that they both had mustaches, and how the dollar had risen over the years since the original Louisiana Purchase. He also joked that since the mayor had just made him an honorary citizen of New Orleans ("You know I am already a Louisiana Colonel," remarked Walt) that maybe he should make the mayor an "honorary dictator of the Magic Kingdom".

When New Orleans Square opened, you could hear the chants and ringing bells of a voodoo queen living off a balcony on the backside of the square near the bathrooms and the train station. It was speculated that it was the sound of the infamous Marie Laveau who practiced voodoo in New Orleans in the 1700s and 1800s. After all, her portrait could be found in both the Pirates of the Caribbean and the Haunted Mansion when those attractions first opened.

Those chants and bells wouldn't have bothered Walt, whose massive personal apartment was being built over the Pirates of the Caribbean ride where his wife Lilly would have been able to easily visit the nearby One-of-A-Kind Shop down below, filled with unique antiques.

At the dedication, Walt said:

> Disneyland has always had a Big River and a Mississippi sternwheeler. It made sense to build a new attraction at the bend of the river, and so New Orleans Square came into being—a New Orleans of a century ago when she was the "Gay Paree" of the American frontier.

Walt's interest in New Orleans goes back to his childhood and his fascination with the big steamboats that journeyed from St. Louis, Missouri, down the Mississippi River to New Orleans.

In fact, one of the things young Walt was looking forward to doing when he returned from France in 1919 was to take a trip down the Mississippi to New Orleans on a homemade raft with his friend, Russell Maas.

The first Mickey Mouse cartoon took place on a steamboat and the song used in the cartoon, "Steamboat Bill", was about a steamboat trying to beat the record of the *Robert E. Lee* steamboat in a race to New Orleans.

On a trip to New Orleans in 1946 with his wife, Lillian, who was an avid antiques collector and loved shopping in the collectibles shops, Walt discovered a small golden bird cage with a singing bird and wondered if this type of technology could be duplicated on a larger scale.

Most research describe this moment as the beginnings of Audio-Animatronics, since Walt gave the bird cage to people at the studio to see if they could duplicate the movements in a human figure.

It was Walt who created the Dixieland at Disneyland event that debuted at the Carnation Gardens on October 1, 1960, and featured big-name entertainers playing Dixieland Jazz, a more frantic, faster version of Chicago Jazz or New York Jazz. Dixieland mixes elements of military bands with street parades and adds syncopation and rhythmic swing.

The following year (September 1961) saw Louis Armstrong join the performances that had been moved on board the *Mark Twain* steamboat. This event was filmed for the "Disneyland After Dark" television episode shown on April 15, 1962, and later released theatrically as a short subject both domestically and overseas.

Armstrong, who was born in New Orleans, performed again in 1962 and from 1964–1967. In 1968, he even recorded an album titled "Disney Songs the Satchmo Way" that brought the flavor of New Orleans to Disney standards.

Louis Prima, who was the voice of King Louie in *The Jungle Book* (1967), was also born in New Orleans.

However, it was not just this special event that showcased the music of New Orleans. The streets of Frontierland echoed with the sounds of the South, performed by groups like Young Men from New Orleans (the gag being they were anything but young) who played there from 1955–1966.

Johnny St. Cyr was the leader of the small group with vocalist Monette Moore, Johnny St. Cyr (banjo), Kid Ory (trombone), Joe Darensbourg/Paul Barn (clarinet), Harvey Brooks (piano), Alton Redd (drums), and Mike Delay (trumpet). He continued to perform at Disneyland until his death in 1966. The Young Men from New Orleans can be seen in the "Disneyland After Dark" special, as well as in clips from various Disneyland parades.

Another group, the Royal Street Bachelors, began performing at Disneyland in 1966. The leader, Jack McVea, was personally hired by Walt Disney. McVea, a famous musician in his own right, having written the song "Open the Door Richard" in 1946, kept the job for 27 years, retiring in 1992.

The original Royal Street Bachelors, who all played string instruments, included Harold Grant (who was replaced by Ernest McLean when Grant died) and Herb Gordy.

Other groups that kept the flair alive at Disneyland were the Side Street Strutters (that had a horn section) and Bayou Brass (that had a Cajun flavor).

Even the Disneyland parades were influenced by New Orleans.

Blaine Kern was known as "Mr. Mardi Gras" because in the 1950s when the Mardi Gras parades were offering only dim shadows of past glories, Kern became an innovator at creating fanciful, outlandish floats that included storybook characters whose heads turned and whose eyes moved.

In 1959, Kern met Walt Disney, who was visiting Mardi Gras in search of new ideas. Walt was quite taken with one of Kern's more inspired creations: an 18-foot-tall King Kong-like gorilla, with five men inside, that walked and made facial expressions.

Remember that this was when Walt was still trying to get a grip on the concept of Audio-Animatronics for his park attractions. Disney showed a clip of Kern's gorilla in an episode of his television program (March 4, 1962) titled "Carnival Time" where Ludwig Von Drake sends Donald Duck to report on the Mardi Gras in New Orleans.

Walt offered Kern a job to work as an Imagineer designing floats for Disneyland, as well as working on other projects. Kern's boss, Darwin Fenner (the son of the "Fenner" in Merrill Lynch, Pierce, Fenner and Smith), convinced Kern to say "no" to Walt.

Fenner reminded Kern of his love for New Orleans, and his passion for Mardi Gras. According to Kern:

> Fenner said, "Son, let me tell you: you stay here in New Orleans, you're gonna be a big fish in a little pond. You go out there, you're gonna be a small fish in a big pond." He said, "Your fortune will be here in the future. Mardi Gras is democratizing; it's opening up to everybody."

Walt's study of Kern's work helped inspire the redesign of Disneyland parades. Today, Kern's son, Barry, carries on the family tradition and is the builder and designer of props and sculptures for Disney and other theme parks.

Of course, the most notable New Orleans influence on a Disney parade was the famous Party Gras parade in Disneyland in 1990 to celebrate 35 Years of Magic. The parade was so impressive that it was shipped out to

Walt Disney World's Magic Kingdom in 1991 to help celebrate that park's 20 nniversary.

Singer, dancers, and stiltwalkers threw beads and a special Party Gras coin to guests and occasionally the parade stopped so the guests could participate. Huge forty-foot-tall floats of Mickey, Minnie, Donald, Goofy, Pluto, and Roger Rabbit rolled down the street and seemed to dwarf the Main Street buildings.

In the early 1980s, Dick Nunis pushed for an expansion of the Shopping Village at Lake Buena Vista (better known today as Disney Springs) by creating a moderately priced resort themed to New Orleans.

The Empress Lilly restaurant would have been a steamboat that had docked to unload its cargo at this riverfront town of New Orleans where the guest rooms would have been "hidden" in buildings resembling a cotton mill or a boatwright's shop. The rooms would be on the upper floors with the bottom floor reserved for shops and restaurants.

This is similar to the design that was eventually used for some resort rooms at Disney's BoardWalk.

A New Orleans-themed resort, named Port Orleans, did appear on Walt Disney World property on May 17, 1991, thanks to Fugleberg Koch Architects of Winter Park, Florida, in collaboration with the Disney Development Company. It was themed to the French Quarter around the mid-1800s and was situated by the Sassagoula River, a man-made Disney waterway named after the Native American word for the Mississippi.

The resort was characteristic of New Orleans' French Quarter, with balconies, wrought-iron railings, cobblestone streets, and courtyards, but with Disney touches like jazz trombone-playing alligators.

The text on the back of the very first postcard released for the resort stated:

> Evoking a bygone era of romance and charm, the hidden courtyards, splashing fountains, and lush gardens of Disney's Port Orleans Resort create a welcome retreat. At the heart of it all is Doubloon Lagoon, where "Scales" the sea serpent invites visitors to make a splash!

Blaine Kern Artists, Inc. were responsible for collecting and creating the many Mardi Gras props, such as the jesters. Some of these props were purchased directly from Mardi Gras warehouses in New Orleans.

On March 1, 2001, the resort officially merged with Dixie Landings and became Port Orleans: French Quarter (with Dixie Landings becoming Port Orleans: Riverside).

Clarence Nash:
The Voice of Donald Duck

Clarence Nash, who did the voice of Donald Duck for more than 50 years, often performed at local schools and events in Glendale, California, where he lived. At these gatherings, he would usually tell the same joke: "My parents wanted me to be a doctor, but they are still proud because I grew up to be the biggest quack in the world."

It always got a big laugh.

Clarence Nash was born December 7, 1904, in Watonga, Oklahoma, roughly three years before Oklahoma became a state. He never grew to be taller than 5 foot 2. Nash was raised on a farm, and for amusement started to imitate the barnyard animals. At age 10, his family moved to a town that is now part of the city of Independence, Missouri.

Nash recalled:

> It was a big thing [at school] for the kids to try to outdo one another imitating animal sounds. By the age of 12, I could do the sounds of dogs, cats, baby chicks, horses, pigs, raccoons, frogs, baby coyotes, and a lot of birds.

It was at this time that he acquired his final childhood pet, a baby billy goat named Mary. Nash had a 1918 photo of her posing with one leg in the air on a wooden box. She died several years later and was laid to rest by a tearful Nash.

"When I got the goat, it was only a couple of days old and I had to feed it with a baby bottle. When I stopped feeding it, it would cry like a frightened little girl," recalled Nash, who was challenged to imitate the unique sound, which he did to the delight of his schoolmates. He also worked at trying to convert the sound into simple words.

For a school talent show when he was 13, Nash recited the poem "Mary Had a Little Lamb" in the voice of the goat, and received wild applause from the audience.

Eventually, he dropped out of school as a teen so he could tour the Midwest as a mandolin player in the Alamo Quintet and an "animal impressionist" on the Redpath Chautauqua and Lyceum circuit. His most popular encore was doing "Mary Had a Little Lamb" in the voice of his billy goat. At the time, he felt that vaudeville would last forever, but was given a shock when it died out in the late 1920s.

Nash married his 18-year-old sweetheart, Margie Seamans, on January 25, 1930, and moved to San Francisco to look for a "normal" job, but in the beginning of the Depression, that was not easy. So the couple moved to Los Angeles where Nash found work doing his animal impressions as part of the KHJ local radio show *The Merry Makers*. That led to a job with the Adohr Milk Company when they heard him perform.

To promote its brand of milk, Adohr hired Nash as Whistling Clarence, the Adohr Bird Man. He would drive a miniature open-topped milk wagon (decorated with games, plugs for the milk, and Clarence's name with painted black bird silhouettes underneath) pulled by a team of miniature horses. He would go to local schoolyards and assemblies to entertain children with his bird calls and animal sounds as well as giving out treats like a small tape measure with Adohr's name on it. He was attired as a standard Adohr milk man.

After two years working for Adohr, as a favor to a friend in San Francisco who missed hearing him on the radio, he made a free appearance on *The Merry Makers*, unaware that Walt Disney was casually listening to the show at that particular moment.

The Disney Studio had just announced that they were looking for someone to provide animal sounds for their animated short cartoons. Nash's friends urged him to audition. Two days after the radio broadcast, he found himself at the studio.

Nash remembered:

> On my way to work one day, I was driving my milk wagon down Hyperion Boulevard. I saw this big billboard picture of Mickey Mouse and it said "Walt Disney Studios, the home of Mickey Mouse." So I pulled over to the curb and decided to go in. I gave the receptionist a circular advertising my work with the milk company and did some bird imitations. I suggested she give it to somebody who might be interested and, two days later, I got a call from animation director Wilfred Jackson to come in and audition.

Nash was still wearing his milk-man outfit because after the audition he had to go back to work. He performed all his animal sounds, his bird calls, and then began to recite "Mary Had a Little Lamb" as his big finish. Jackson stopped him for a moment. He switched on the intercom which went directly to Walt's office and then asked Nash to continue.

Within moments, Walt rushed into Jackson's office and said excitedly, "That's our talking duck!"

Nash was smart enough not to say it was the voice of a goat. Walt was thinking of doing a cartoon with a girl duck, but it hadn't yet jelled as a story.

As he was leaving, Nash ran into Disney storyman Ted Osborne, who had produced the recent episode of the radio show featuring Nash, and said, "Hey, Walt heard you that night and he was going to look you up." Apparently, during a break at a late-night story meeting, Osborne had turned on the show to see how it was going, and Walt heard Nash.

Nash was not hired, but put on a retainer for a year, so he kept his job with the dairy. His first actual voice work for Disney was supplying bird sounds in the Mickey Mouse short *The Pet Store* (1933) and he kept being tossed similar little bits.

At that time Walt's former partner, Ub Iwerks, had left to start his own animation studio and had also heard of Nash's expertise. He asked Nash to come in and audition to do a voice or voices for his new cartoon short in production, *The Little Red Hen*, based on the children's story of a hen looking for help to plant, harvest, and grind her corn.

Nash wisely phoned to ask Walt's permission, but Walt was unavailable. Ironically, Walt also had a version of the same story, titled *The Wise Little Hen,* in development and was planning to have Nash do the voice of the duck.

At the time, those working at the Disney Studio thought Donald's co-star Peter Pig was going to be the big breakout star of the short. There had also been plans to include a character named Tom Turkey, but he was cut before the short went into production.

Walt called Nash back roughly a dozen times, but couldn't reach him since he was out on the streets doing his job. He finally left a message for Nash not to do anything for the Iwerks cartoon.

Nash got called into a story meeting for *The Wise Little Hen* and met Disney storyman Pinto Colvig who told him:

> Walt told me about your calling about your going over to Iwerks. He said, "I like that loyalty in that guy, I'm going to put him on the payroll."

After the story meeting, Walt asked Nash to come to his office to talk about revising the retainer agreement so that Nash could still work for Adohr while doing voices at Disney.

As Nash told animation historian Milt Gray

> I told Walt, "I'd rather work for just one organization, I don't want to work for two people." "OK, we'll start you at the first of the year." "Walt, if you don't mind, I'd like to come to work sooner."

On December 2, 1933, Nash became Disney's 125th employee. He was earning the same amount he made at Adohr: $35 a week.

Nash could not put in enough hours just doing voices to justify that full-time salary, so he often found himself temporarily in side jobs, from accepting artist portfolios and processing them to being a chauffeur for visiting celebrities. His wife's reaction to him being hired to do the voice of a duck was, "That's nice, but it probably won't last."

Nash recalled:

> Donald lasted a lot longer than I thought he would. I figured he was just like any other cartoon character who would eventually run his course. It never occurred to me he would last over 50 years.

Disney's *The Wise Little Hen* was released on June 9, 1934 (now Donald Duck's official birth date). Iwerks' *The Little Red Hen* was released February 16, 1934, as part of the ComiColor series, and is rarely known or seen today.

Nash was not immediately known to the public as the voice of Donald Duck. Walt always felt that the voice was just one of many elements in a character and so actively tried not to publicize any particular vocal artist. He felt that the lines the storyman wrote, the design of the character by the animator, and even how other animators made the character move and react were equally important.

Nash said:

> It was all right with me that people didn't know who I was, but I was happy when they eventually did find out. In the early days, Walt didn't want us voices to have any publicity. I went along with his wishes, but one time my name got out in the newspapers. Walt and I had a big argument over it, but, when I left his office, I wasn't upset. Walt was a very fair man. I ended up with a raise.
>
> Before Walt came up with the idea, I had never even thought of being angry or laughing in that voice. But the more I learned to use it, the more it developed. Walt believed it was important for Donald to have a strong personality so he would seem alive.

Donald's voice was known as a "stunt voice" because it was not created using the familiar voice placements or combinations that most voice actors use to create a voice.

However, this did not prevent Hanna-Barbera from trying to duplicate that ducky sound both in several of the Tom and Jerry theatrical cartoons with *Little Quacker* (voiced by "Red" Coffey, the stage name of Merle Coffman) and later on in their Yogi Bear television series with *Yakky Doodle* (voiced originally by Jimmy Weldon, who was a local Los Angeles children's show host with a ventriloquist doll named Webster Webfoot who also spoke like Donald Duck).

These imitations often led to confusion.

In 1977, when I talked with him, Clarence Nash was not happy:

> Everybody thinks Mel Blanc is Donald Duck! He's not. I'm Donald Duck. We've had some problems with people who say they're the

"original Donald Duck" and we've even had some problems with them at the Disney Studio in the past. Every once in a while, we hear that I died and we get Christmas cards saying they're sorry I passed away during the year.

Donald cartoons are shown in dozens of countries and are usually dubbed rather than subtitled. In the beginning, for foreign releases, Donald's voice was dubbed by Nash into a foreign language. The words were written out phonetically for him, but generally, Donald had very little spoken dialog.

Nash stated:

> I had to learn to quack in Spanish, Portuguese, French, Italian, German, Dutch, Swedish, and even Chinese. There were, however, foreign language coaches who helped me. I listened through earphones to the English dialogue, and I'd match the length and mood of the dialogue in that other language. It was critical to get everything down pat so they never had to re-animate. It had to seem like the language came out smoothly and matched the mouth movements of Donald.

Disney Legend Carl Barks recalled at NEWCON 1976:

> Donald evolved out of Ducky Nash's way of saying things in duck talk. He would quack, quack, quack, and blow words out of the side of his mouth or something and that created Donald. They just wanted some character that would fit the crazy sound that Clarence Nash was making. He used to come in on the story meetings.
>
> We had a lot of dialogue that he had to practice. And it would determine, sometimes, what the dialogue would finally be, whether or not he could say the words. Of course, none of us could understand him even when he said, "Well, I said that all right."

When I interviewed Donald Duck director Jack Hannah in 1979, he said:

> Clarence was always nice to work with. He did many little side voices, such as meowing cats or miscellaneous characters. One problem we always had was the understanding of Duck lines. He was great when he lost his temper and all of that.
>
> However we had to pantomime pretty well in the drawing what the Duck was thinking or doing because if you tried to get over a gag or a line of dialog with understanding, you were in trouble using that voice. One thing I'll say about Clarence Nash: he was a hard worker and I actually thought a couple of times he was going to faint on me. The blood would come to his face on these wild tantrums especially where they were prolonged.
>
> It was the bane of my existence to get that voice understood! It was aggravating as hell to do a picture with dialogue and not be able to understand the main character. But he did have a variety of moods and you could get over with strong poses what he was trying to tell you.

I got some old acetates of a television show I made and I noticed that Jimmie Dodd says, "And now Donald we're going to take you around the world." The Duck asks, "Around the world?" Jimmie replies, "Yes, that's right. Around the world." We did it that way to be damn sure you could understand what was being said. Once a human said it, then you could understand the way the Duck said it. We did that in some of the cartoons, as well. If you heard the line repeated by a straight voice, it made it easier to understand the Duck.

We always did a minus dialogue track whenever we did a cartoon so they could do foreign voices and fill them in for foreign release. Jack Cutting was in charge of the foreign department and he made sure the foreign voices were done and I never had anything to do with any of that.

Well, not long ago, I talked with Clarence Nash on the phone. I mentioned something like, "Well, on those foreign voices, that was one good thing. You didn't have to redo the Duck in different languages." But Clarence was very proud of the fact that anybody could understand him doing the Duck and he replied, "Oh, yes, I do Spanish. Listen to this."

And he did it in Spanish over the phone and he did a couple of other languages and to me it still sounded like the same old English you couldn't understand. But, Clarence, knowing what he was supposed to be saying, naturally thought everyone else could understand him.

On the "Donald's Silver Anniversary" episode (November 1960) of the weekly Disney television show, Walt Disney stated:

But of all Donald's accomplishments, we're the most proud of his efforts in spreading good will throughout the world. You might say Donald speaks a universal language. That is to say, that no one can understand what he says in any language, but the whole world laughs at him.

Besides Donald, Nash often supplied bird calls and animal sounds for Disney characters, from Uncle Remus' bluebird in *Song of the South* to the meows for little Figaro the kitten in a handful of short cartoons to the earliest voices for Huey, Dewey, and Louie.

Nash was actively involved in 1984 with the 50th birthday celebration for Donald Duck, touring the country, giving interviews and appearing at special events.

In order not to spoil the festivities, Nash did not let people know he was suffering from leukemia. In what turned out to be his final public appearance, he went back to his hometown of Watonga, Oklahoma, on December 7, 1984, where Governor George Nigh declared that day to be Clarence Nash Day throughout Oklahoma. Watonga renamed a street Clarence Nash Boulevard in his honor.

Nash was unfortunately too ill to ride on the City of Glendale Tournament of Roses float on January 1, 1985, which was themed to Donald Duck. He died on February 20, 1985 at the age of 80. Nash's tombstone in the San Fernando Mission Cemetery in Mission Hills, California, is shared with his wife (who died in 1993) and has a carving of Donald and Daisy Duck holding hands.

Today, animator Tony Anselmo does Donald's voice, as he has done since Nash's passing.

One of my favorite episodes of the Disney weekly television show was "A Day in the Life of Donald Duck" (February 1956) where the real life Clarence Nash meets an angry animated Donald Duck in Donald's office at the Disney Studio.

Donald Duck [handing Clarence Nash a fan letter]: Read this.

Clarence Nash [reading letter]: Can't understand me?

Donald Duck: ME, you fathead!

Clarence Nash: Oh, can't understand you. Must be that fat beak of yours. [Nash shapes his hand like a duck's beak and opens and closes it.] Can't mouth the words right.

Donald Duck: Why, you pigeon brain, you just don't articulate!

Clarence Nash: Oh, yeah? Well, listen to this... [In Donald's voice:] Peter Piper picked a peck of pickled peppers!

Donald Duck: Phooey! Can't understand you myself!

Acknowledgments

As always, I would like to acknowledge not only the people who directly helped me with this specific book, but those who have inspired or supported me over the years. Nobody does anything alone. We all stand on the shoulders of others.

I would like to acknowledge and thank all the people who have bought my Disney history books because their support has allowed this book to be published.

This book would not have been possible without the skills and encouragement of publisher Bob McLain and his Theme Park Press.

Thanks to my brothers, Michael and Chris, and their families, including their children—Amber, Keith, Autumn, and Story—who never really understood what their uncle does or why he does it. Also, my grand-nieces Skylar, Shea, and Sidnee, who are already Disney princesses and act accordingly.

Many thanks to Didier Ghez, Leonard Maltin, JB Kaufman, Werner Weiss, Sam Genawey, Michael Barrier, Dave Smith, Kim Eggink, Diane Disney Miller, John Cawley, Bill Iadonisi, Chris Strodder, Marion Quarmby, Sarah Tabac, Malcolm and Mary Joseph (and their children: Melissa, Megan, Rachel, Nicole, Richard);

Tom and Marina Stern, Jerry and Liz Edwards, Lonnie Hicks, Kirk Bowman, Jeff Kurtti, Ryan N. March, Brad Anderson, Kaye Bundey, Brian Sibley, David Gerstein, Alycia Leach, Lucas Seastrom;

Todd James Pierce, Paul Anderson, Betty Bjerrum, Jerry Beck, Dr. Mark Round, Dave Mruz, Tracy M. Barnes, Sarah Pate, Tamysen Hall, Evlyn Gould, Bruce Gordon, David Mumford, Randy Bright, Jack and Leon Janzen, Arlen Miller, David Lesjak, Howard Kalov, Dana Gabbard, Heather Sweeney .

And sadly some people that I have foolishly forgotten for the moment. Their kindness and generosity, like those names listed here, have lightened my journey through life and made this book possible. I hope all of you, both acknowledged and temporarily missing, live happily ever after and enjoy this book.

About the Author

Jim Korkis is an internationally respected Disney historian who has written hundreds of articles about all things Disney for over three decades. He is also an award-winning teacher, professional actor and magician, and author of several books.

Jim grew up in Glendale, California, right next to Burbank, the home of the Disney studio.

As a teenager, Jim got a chance to meet Disney animators and Imagineers who lived nearby and began writing about them for local newspapers. Over the decades, Jim pursued a teaching career as well as a performing career, but was still active in writing about Disney for various magazines.

In 1995, he relocated to Orlando, Florida, to take care of his ailing parents. He got a job doing magic and making balloon animals for guests at Pleasure Island. Within a month, he was moved over to the Magic Kingdom, where he "assisted in the portrayal of" Prospector Pat in Frontierland as well as Merlin the Magician in Fantasyland for the Sword in the Stone ceremony.

In 1996, he became a full-time salaried animation instructor at the Disney Institute where he taught every animation class, including several that only he taught. He also instructed classes on animation history and improvisational acting techniques for the interns at Disney Feature Animation Florida. As the Disney Institute re-organized, Jim joined Disney Adult Discoveries, the group that researched, wrote, and facilitated backstage tours and programs for Disney guests and Disneyana conventions.

Eventually, Jim moved to Epcot where he was a coordinator with College and International Programs and then a coordinator for the Epcot Disney Learning Center. During his time at Epcot, Jim researched, wrote, and facilitated over two hundred different presentations on Disney history for Disney cast members and Disney's corporate clients including Feld Entertainment, Kodak, Blue Cross, Toys "R" Us, and Military Sales.

Jim was the off-camera announcer for the syndicated television series *Secrets of the Animal Kingdom*; wrote articles for Disney publications like *Disney Adventures, Disney Files* (DVC), *Sketches,* and *Disney Insider.* He worked on special projects like writing text for WDW trading cards, as the on-camera host for the 100 Years of Magic Vacation planning video, as facilitator with the Disney Crew puppet show, and countless other credits, such as assisting Disney Cruise Line, WDW Travel Company, Imagineering, and Disney Design Group with Disney historical material. As a result, Jim was the recipient of the prestigious Disney award, Partners in Excellence, in 2004. (Jim is not currently an employee of the Disney Company.)

Several websites feature Jim's essays about Disney history:

- MousePlanet.com
- AllEars.net
- Yesterland.com
- CartoonResearch.com
- WDWRadio.com
- YourFirstVisit.net

To read more stories by Jim Korkis about Disney history, please purchase his other books, all available from Theme Park Press (ThemeParkPress.com).

The Vault of Walt

Disney History at Its Best

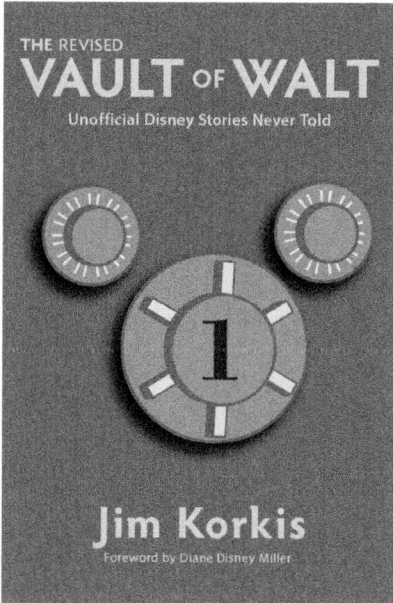

THE REVISED
VAULT OF **WALT**
Unofficial Disney Stories Never Told

1

Jim Korkis
Foreword by Diane Disney Miller

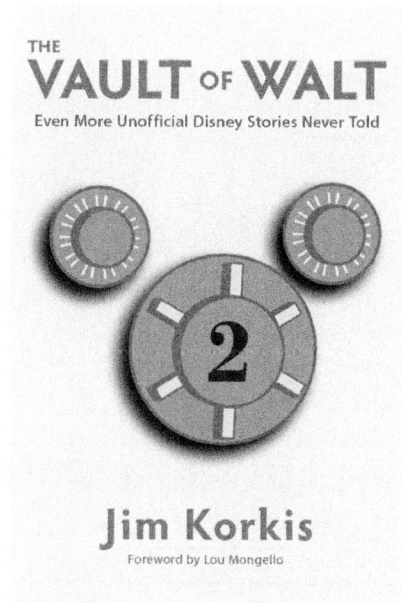

THE
VAULT OF **WALT**
Even More Unofficial Disney Stories Never Told

2

Jim Korkis
Foreword by Lou Mongello

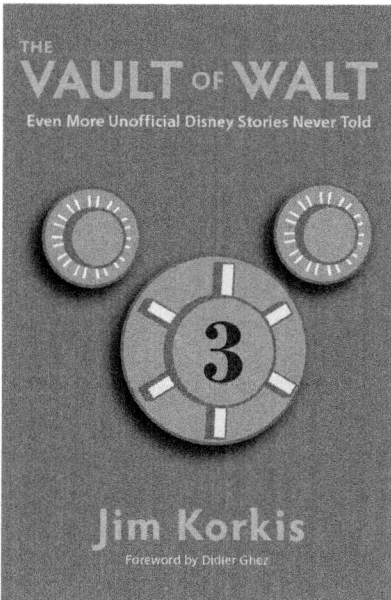

THE
VAULT OF **WALT**
Even More Unofficial Disney Stories Never Told

3

Jim Korkis
Foreword by Didier Ghez

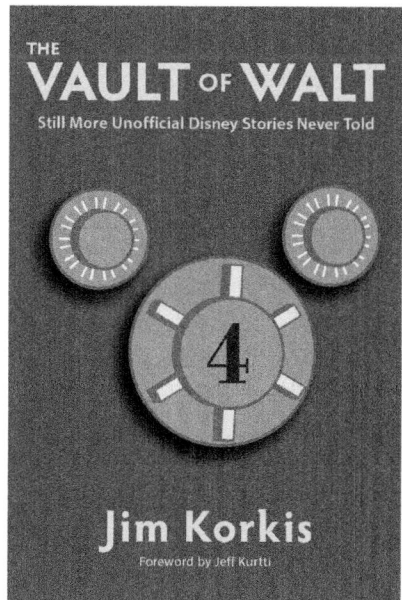

THE
VAULT OF **WALT**
Still More Unofficial Disney Stories Never Told

4

Jim Korkis
Foreword by Jeff Kurtti

More Books from Theme Park Press

Theme Park Press publishes dozens of books each year for Disney fans and for general and academic audiences. Here are just a few of our titles. For the complete catalog, including book descriptions and excerpts, please visit:

ThemeParkPress.com

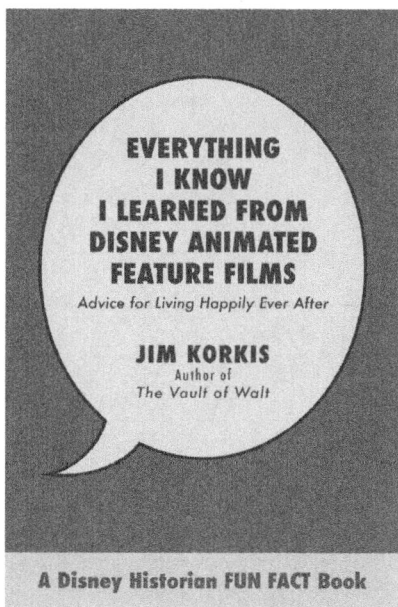

EVERYTHING I KNOW I LEARNED FROM DISNEY ANIMATED FEATURE FILMS

Advice for Living Happily Ever After

JIM KORKIS
Author of
The Vault of Walt

A Disney Historian FUN FACT Book

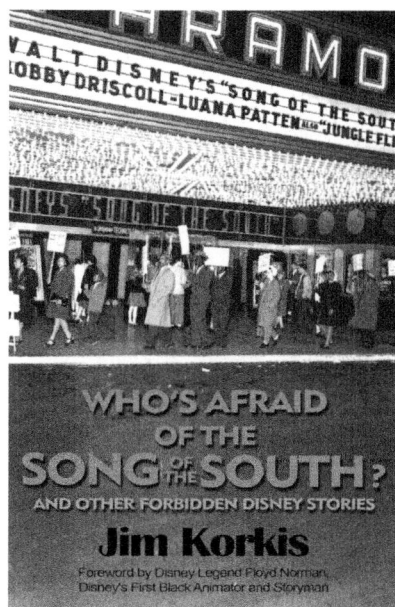

WALT DISNEY'S "SONG OF THE SOUT
OBBY DRISCOLL · LUANA PATTEN ALSO "JUNGLE FLI

WHO'S AFRAID OF THE SONG OF THE SOUTH?
AND OTHER FORBIDDEN DISNEY STORIES

Jim Korkis

Foreword by Disney Legend Floyd Norman,
Disney's First Black Animator and Storyman

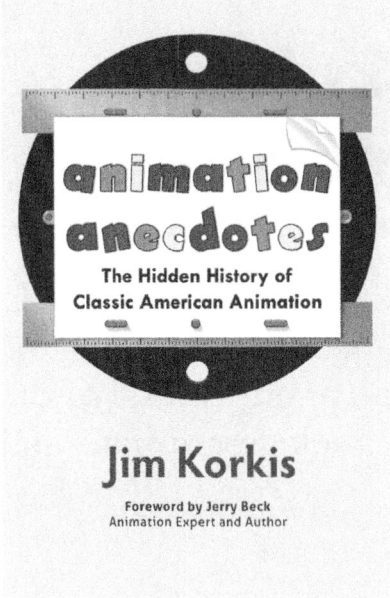

animation anecdotes
The Hidden History of Classic American Animation

Jim Korkis

Foreword by Jerry Beck
Animation Expert and Author

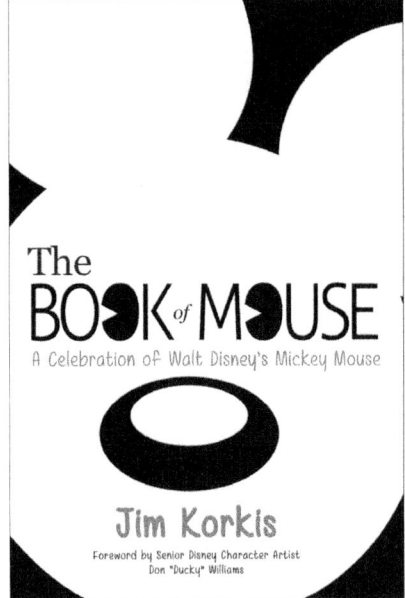

The BOOK of MOUSE
A Celebration of Walt Disney's Mickey Mouse

Jim Korkis

Foreword by Senior Disney Character Artist
Don "Ducky" Williams

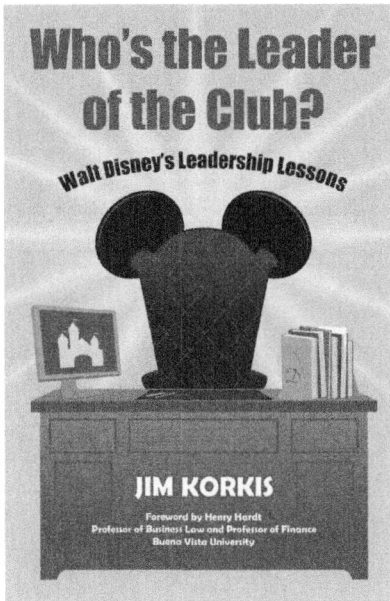

Who's the Leader of the Club?
Walt Disney's Leadership Lessons

JIM KORKIS

Foreword by Henry Hardt
Professor of Business Law and Professor of Finance
Buena Vista University

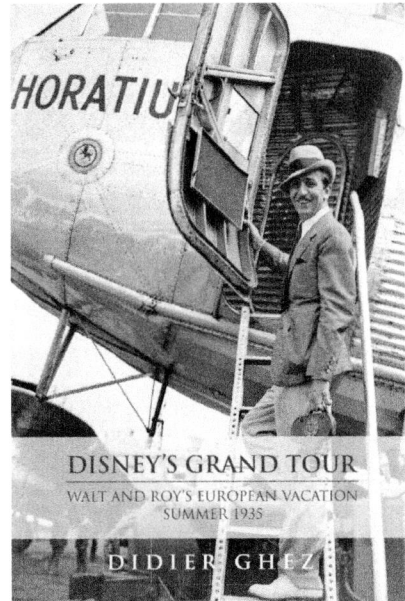

HORATIU

DISNEY'S GRAND TOUR
WALT AND ROY'S EUROPEAN VACATION
SUMMER 1935

DIDIER GHEZ

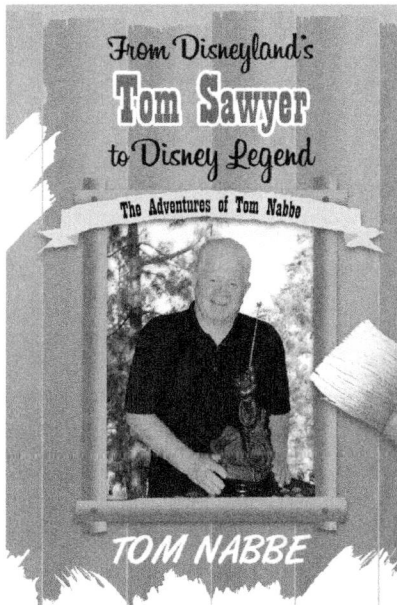

From Disneyland's
Tom Sawyer
to Disney Legend

The Adventures of Tom Nabbe

TOM NABBE

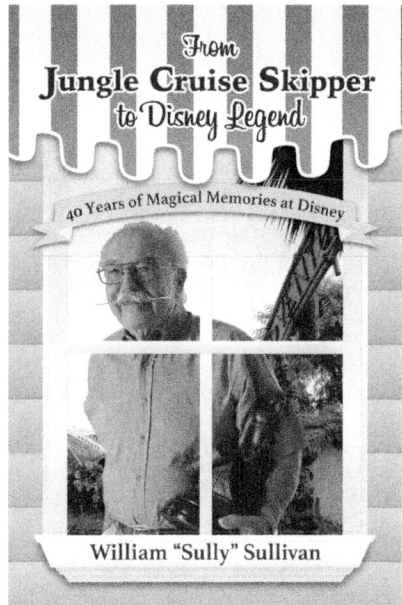

From
Jungle Cruise Skipper
to Disney Legend

40 Years of Magical Memories at Disney

William "Sully" Sullivan

IT'S A
CRAZY BUSINESS

The
GOOFY
Life of a
Disney
Legend

Pinto Colvig

Edited and Introduced by Todd James Pierce

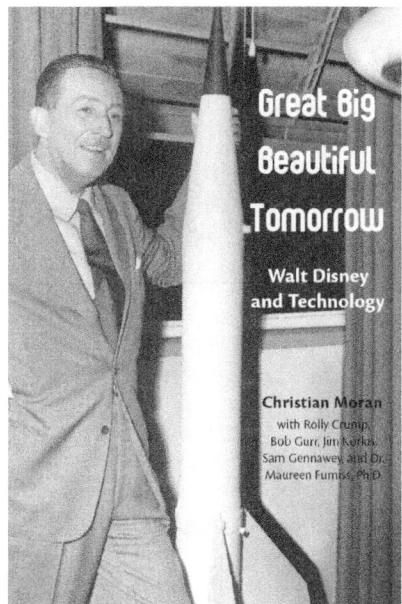

Great Big
Beautiful
Tomorrow

Walt Disney
and Technology

Christian Moran
with Rolly Crump,
Bob Gurr, Jim Korkis,
Sam Gennawey, and Dr.
Maureen Furniss, PhD

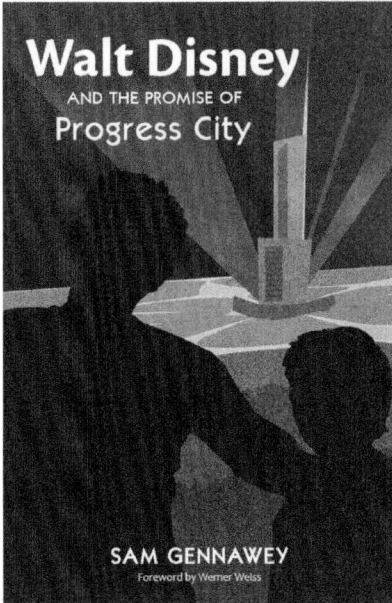

Walt Disney
AND THE PROMISE OF
Progress City

SAM GENNAWEY
Foreword by Werner Weiss

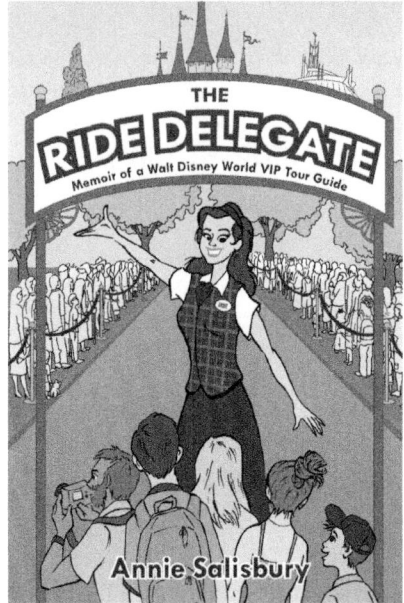

THE
RIDE DELEGATE
Memoir of a Walt Disney World VIP Tour Guide

Annie Salisbury

Disneyland
SECRETS

GAVIN DOYLE

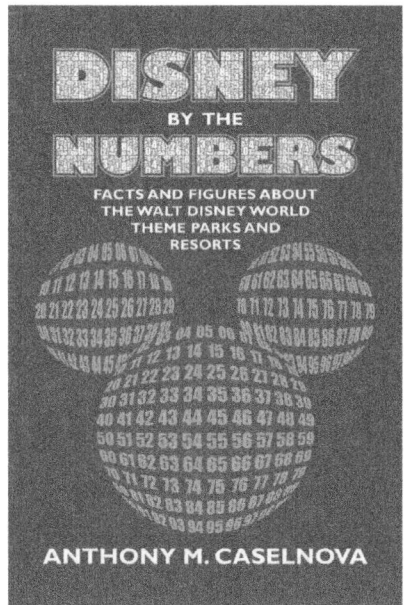

DISNEY
BY THE
NUMBERS

FACTS AND FIGURES ABOUT
THE WALT DISNEY WORLD
THEME PARKS AND
RESORTS

ANTHONY M. CASELNOVA